HYDROPONICS:
An Unconventional Step-by-step Guide For Beginners on How to Building and Managing a High Quality Hydroponic System

MATTHEW GARDEN

Copyright ©2020 Matthew Garden

All rights reserved.

Table of Contents

INTRODUCTION
CHAPTER ONE 2
CHAPTER TWO 17
CHAPTER THREE 46
CHAPTER FOUR 88
CHAPTER FIVE 94
CHAPTER SIX 103
CHAPTER SEVEN 108
CHAPTER EIGHT 117
CHAPTER NINE 132
CONCLUSION

INTRODUCTION

Welcome!

Welcome to this guide on hydroponics. If you are a beginner, you have already done the first step for reaching the knowledge necessary for your goal: build your own hydroponically system. Congratulations! With the help of this book you will know both the science of hydroponics and its practical applications, and see that my book is not only a way of preventing the purchase of costly hydroponic systems; it is also a way to create a beautiful garden that is ideally suited to your needs, the same garden you have always dreamed of. Offering hydroponic gardens building guides that vary from easy to complex, this book will be suitable for almost any setting or application.

The build guides provide plenty of options to customize the design, advices for what regards crop selection and budget. Additionally, this book provides useful suggestions for seed varieties that can save time and money for new hydroponic gardeners that could easily have been lost on unsuited crop choices. Learn from this unconventional guide and prevent the costly mistakes that new hydroponic growers usually produce.

Thanks for choosing this book, I'd really love to hear your thoughts. Good reading. And remember:

The more you are conscious, the stronger you are.

WHAT IS HYDROPONICS?

I know that you probably know what hydroponics consists in, but let me explain it more specifically. Hydroponics, to put it simply, is a technique that allow plants to grow without soil. How is it possible? It's easy, because plants grows in water. Is it powerful? Oh yes! Most people believe that soil is important to plant growth, but now you are not part of this people.

The different soil functions can be recreated with the aid of other materials. Soil provides plant protection as it produces a physical foundation that enables the roots to grasp. Tall trees would not be able to stand upright on a windy day without a strong grip in the soil. In a hydroponic system, a variety of materials and trellis structures can mimic the physical support offered by the soil. Soil also contains vital nutrients for growing plants. Nevertheless, these same nutrients can be supplied using alternative methods. Hydroponic systems have water-soluble nutrients, both organic and conventional. Soil may also provide a habitat for important microbial species establishing beneficial connections with plant roots. In a hydroponic system, these very microbes can live and thrive.

CHAPTER ONE

THE HYDROPONICS SYSTEM

Here we discuss the different parts of the system and why not, I will tell you the history of hydroponics. The critical elements the system is made of are:

- Nutrient solution
- Gas exchange
- Lighting
- Growth / Support material such as perlite, rock-wool, coconut coir, etc.

It is said that the ancient Babylonians, the Aztecs and the Egyptians used Hydroponic irrigation to grow plants; one such example is the popular' Hanging Gardens of Babylon.' Hydroponics studies started in 1936, University of California's Dr. Gericke in Los Angeles, succeeded in growing tomatoes in water culture. Hydroponics was utilized around World War II to grow plants in non-arable areas. It's not until around this time: 1970s, however, that farmers and gardeners started to show interest in Hydroponics. But, still, to this date, the full potential of Hydroponics is not exploited; the concept is still unknown to many gardeners.

Hydroponic systems are commonly used to cultivate a variety of plants such as tomato, green pepper, cabbage, basil etc. The Hydroponic cultivation method is seen as a potential solution to the world's hunger problems. The system's beauty lies in its simplicity. To grow plants, you can use your garage, the rooftop or any room available. Continuous research and development have led to the sustainable and long-lasting use of PVC material; Middle Eastern countries where water is a scarce resource follow Hydroponic cultivation system. Huge buildings are designed to assist in crop production with desalination systems. The future of Hydroponic cultivation is getting brighter by the day, and it would not be shocking to see this cultivation system becoming the standard!

HYDROPONIC GROWING ADVANTAGES

1. SOIL NOT NECESSARY: Gardening is mostly seen as an occupation limited to those lucky enough to have a lawn. Hydroponics significantly improves the possibilities of planting for people who don't own lawn or those with lawns that have poorly prepared soil for edible crops. Combined with indoor growing techniques, hydroponics offers gardeners even more options by extending the available garden area to almost anywhere inside your house. Wonderful.

2. FASTER CROP GROWTH: Plants never reach their maximum growth in soils. It's a matter of fact. Indeed, their growth is almost always hampered by any limiting factor. Through soil, the plant roots continue to look for nutrients that are often unevenly distributed and likely unavailable as they are attached to specific particles of the soil. Some nutrients are not usable because the soil's microbes have yet to break down the source of nutrients (e.g., manure) into a shape available to the plant's roots.

Another factor that limit the plant production is the amount of water: it have not to be excessive, not even too little. Too much water will reduce the oxygen available to the roots and hinder the biological processes required to take up nutrients and water from the roots. Hydroponics bathes the roots in an appropriate mix of vital nutrients with a water-oxygen balance. Thanks to hydroponics and indoor growing techniques, many of the restrictions on the future growth of a plant can be removed.

3. LESS SPACE OCCUPIED: In the search for water and nutrients, a plant has to spread its roots far and wide. By removing the need to find water and nutrients for the plant roots, the spacing of plants is restricted only by the area required for the plant canopy.

4. LESS CONSTRAINT ON GROWING SEASON: It goes without saying that growing indoors helps gardeners to prolong the growing season. More clearly, but when put outdoors, hydroponics directly will prolong the growing season. The temperature of the roots of a plant is always more important for its survival than the temperature of the leaves. Winter crops can be grown at 100 ° F if the root temperature is kept close to 65 °-75 ° F in an optimum range. Crops which prefer warm temperatures in cold climates can also be cultivated by increasing the temperature of the root zone. Hydroponics increases the ability to change the temperature of the root zone with precision. A hydroponic gardener can raise or decrease water temperature and boost crop growth with the use of heaters, chillers, or simple practices such as burying a hydroponic reservoir.

5. CAN BE USED IN ANY LOCATION: Hydroponics helps gardeners to thrive in places where the soil is not of quality. Hydroponics also allows gardeners to thrive in areas that would be unsuitable for crops due to inhospitable climate or insufficient access to water. In deserts, one of the greatest opportunities for hydroponics is to expand. Deserts also have a wonderful environment for growing crops, with plenty of light and little presence of pests, but access to water is minimal. Hydroponics uses far less water than conventional methods, which can make farming a viable choice in deserts. The main method usually practiced to grow plants in space is hydroponics, too. Most crops have been cultivated in space using hydroponic techniques, including the lettuce.

6. DON'T USE TOO MUCH WATER: Hydroponics takes little water as you may (want to) reuse any irrigation water that is not taken directly from the crop. Most of the water in the soil is lost to evaporation and runoff. For hydroponics, by covering the water tank, the evaporation can be minimized or removed, and all runoff water is stored for reuse.

7: NO WEEDING AND NO HERBICIDES: On first it may seem like a minor point, but after a season of pulling weeds from the garden, most typical soil gardeners would love to have spent that time doing something more enjoyable, like preparing their harvest dishes. Hydroponic farmers don't need to buy herbicides either. Furthermore, if a wind inadvertently blows herbicide into your garden and injures or destroys your precious plants, hydroponic growers will never face possible crop damage from herbicide drift. Leafy vegetables can be grown hydroponically in outdoor space

8. CAN ERADICATE OR CANCEL THE NEED FOR PESTICIDES: Hydroponic gardens, particularly outdoor and greenhouse gardens, are rarely free of pests, but hydroponics can reduce pest pressure. Hydroponic gardens offer less insect hiding places that burrow into the soil or hide in rotting plant debris. If hydroponics is paired with indoor growing techniques, if the gardener uses proactive pest control techniques, it is possible to have a fully pest-free garden. Preventive pest control methods are described in the Equipment section of the chapter Equipment for Rising Indoors.

9. MAY REDUCE OR ELIMINATE AGRICULTURAL RUNOFF: The runoff in a typical garden is difficult to handle. The gardener will fertilize the garden, and much of the fertilizer is washed away by a rainstorm the next week. It's also likely that regular irrigation would take the nutrients away. Use advanced hydroponic techniques zero runoff is possible. It is a best-suited technique for experienced hydroponic growers as it requires advanced water processing, chemistry and detailed knowledge of the basic nutrient needs of a crop. For home hydroponic gardeners, the nutrient solution in the hydroponic system is usually flushed or dumped every few weeks to prevent possible nutrient disorders in the crop caused by an unbalance of nutrients. Plants don't absorb all the nutrients at the same rate, so some accumulate over time, and some get deficient. Periodic flushes, or adjustments in the nutrient solution, help reset the system to ensure that the crop has access to sufficient nutrient balance. Nevertheless, this runoff doesn't just need to be washed down; most hydroponic gardeners use this water for their outdoor garden or potted plants. A conventional garden based on the soil is a perfect accompaniment to a hydroponic garden.

10. ABILITY TO CONTROL NUTRIENT CONTENT: One of the most common myths concerning hydroponics is that hydroponic goods have a lower nutrient density than soil-based crops, since hydroponic crops are based in water. Several experiments comparing the nutrient content of hydroponic and soil-grown items have been performed, and the findings are fairly mixed. There are so many factors influencing a crop's nutrient content, and while fertilizer does play a role in which nutrients are available, the climate has a huge role in which the plant actually absorbs nutrients. Antioxidant quality can be influenced by light intensity and different light colours. Antioxidant production may be impaired by stress due to irrigation activity. Temperature can affect the concentration of sugars. There is a long list of factors influencing the nutrients in a crop, but these vegetables are nutritious overall. The variations are very minute, and when you eat a vegetable, it is hard to go wrong. Almost all plants will display clear signs of nutrient deficiency if their nutrient density is substantially lower than normal levels, and if the plant looks healthy, it is more than likely to have a nutrient profile equivalent to a similar plant, regardless of the environment it was grown in. Having said that, there are several special methods used by hydroponic growers to control their crops. Most commercial hydroponic tomato growers deliberately stress their plants with high levels of nutrients at key stages of their growth to cause an increase in tomato sugar content. Growers should spike the nutrients to cause the rise in sugar and then reduce the nutrients to a normal level to keep healthy growth going. For lettuce, the Oizumi Yasaikobo Co., Ltd. has developed a method for growing low-potassium lettuce using hydroponic methods in Chichibu City, Japan. For clients suffering from kidney failure who are being treated with dialysis, the farm grows these specialty lettuces and is prohibited from eating vegetables with a high potassium content. This initiative to grow produce with a custom nutrient content is one of many similar projects being built around the world as growers acquire the ability to monitor every aspect of a growing environment with precision.

11. IMPROVED ABILITY TO CONTROL CROP GROWTH FOR DIFFERENT CHARACTERISTICS: It is not only possible to manipulate the nutrient content, but it is also possible to manipulate other characteristics such as leaf size, leaf color, root size and plant height when hydroponics are combined with indoor growth. Indoor gardeners can use different luminous colors to induce similar characteristics. Blue light's utilization to grow more compact greens indoors is a common practice to reduce the vertical space needed for a crop.

12. CLEAN AND LOW MESS: Gardening can be messy. That's not terrible but it's not always perfect. The International Space Station is the most extreme example. A floating layer of soil around the delicate equipment will be a tragedy. The advantage of soilless cultivation is a cleaner crop for those of us who do not grow plants in vacuum. Hydroponically grown crops also require little to no washing. Hydroponic gardens can be a perfect way to introduce children to plants in a classroom or home without bringing a major muddy mess into the future. One of my favourite kid-friendly systems is the hydroponic fairy garden listed in the Hydroponic Growing Systems section of the Media Beds site.

13. IT IS EASIER AND SAVES STREES THAN GROWING IN THE SOIL: Fertilizers are easy to use, easy to automate, and weeding is just a couple of the reasons why hydroponic gardening can be much easier than conventional methods. Hydroponics may seem daunting to beginners, but they are hooked after one crop or two of the most hydroponic gardeners. Easy to learn and reproduce results, Hydroponic crops grow increasingly, enabling farmers to gain more experience in a shorter time span. Experience is the best instructor, and faster-growing crops allow hydroponic growers to learn quickly. When a grower works out the best formula for that setting and the chosen crop, the process can be easily repeated. Hydroponics gives the farmers the ability to reproduce the exact available nutrients and the duration of irrigation. Through pairing hydroponics with indoor growing techniques, farmers are increasing their power even further. Without the seasonal and annual variations faced by conventional gardeners, indoor gardeners may mimic light intensity, light period, temperature, humidity, carbon dioxide levels and airflow to grow predictable crops year-round.

15. IMPROVES ABILITY TO CONTROL SOIL-BORNE PATHOGENS, INCLUDING ROOT ROTS AND BACTERIAL WILTS: Are soil-borne some of the most destructive plant pathogens. Every grower who has fought root rot or bacterial wilt in a conventional garden knows that eradicating the problem is very difficult. Some of these pathogens hide until the conditions are right in the soil, and then they spring into action. For hydroponics, if there is a case of a soil-borne pathogen, the gardener should clean out the hydroponic system entirely. That allows the gardener to extract the old crop quickly, clean and sterilize the system, and then start a new crop.

16. POTENTIAL REDUCES OF CONTAMINATION CROPS:

Many cases of the national food-borne disease have been traced back to manure. Animal waste, one of the conventional farms ' key nutrient sources, is a possible source of harmful pathogens like E. Coli, Listeria, and Salmonella, unless prepared correctly before use. The problem is that the farmer doesn't apply all the manures present in agricultural fields. An E. 2011. Coli outbreak in Oregon was believed to have been caused by deer feces found on a suspicious farm. In hydroponics, it is rare to see any manure-derived fertilizers, and wildlife pollution is also unusual, as most hydroponic farms are in managed environments that exclude wildlife. Heavy metals found in soil or wastewater sources are another possible cause of pollution. Hydroponic growers can easily filter their source of water to reduce heavy metals, but it can be very difficult to extract heavy metals from the soil. Research indicates that edible crops will absorb heavy metals which could lead to slow, heavy metal poisoning if grown using polluted soil or contaminated water sources.

KEY FEATURES OF A HYDROPONIC

The hydroponic system is very basic. To build one, you'll need some kind of waterproof reservoir to hold the nutrient solution, and the plants themselves in some cases. You will need a growing area too, which is where the plants will live. The size and type of area that grows determines the types of plants that you can grow and how much they yield. In most cases, you will also need the lighting and ventilation systems. Finally, growing media is necessary to store and release nutrients to the roots of plants.

THE TANK

Most hydroponic systems have a tank filled with a solution of nutrients, a combination of fertilizer and water. There are several choices in hydroponic gardens for the nutrient sources. With a wide range of plants, most nutrient solutions can be used, or they can be catered with particular crops. It is as convenient to feed plants in a hydroponic garden as to make iced tea from concentrates. Simply mix the concentrate in the powder or liquid, stir and finish! Reservoirs can be constructed by repurposing traditional household objects, such as storage totes; they can be constructed with wood and a plastic liner, or they can be purchased. Reservoirs may be as easy as a bottle of plastic or of glass.

THE GROWING AREA

The growing area can be changed to grow virtually every plant in a hydroponic garden. The hydroponic gardeners can establish optimal growing conditions for any crop they want by adjusting irrigation frequency, pot/tray size, substratum, and climate. Many crops are more realistic than others; hydroponic wheat and maize are feasible, for example, but they also require large areas for proper pollination, and with a capital-intensive growing process, the economic value of their yield is small and hard to justify. However, most hydroponic gardeners consider many advantages over conventional growing methods when devoting their growing area to vegetables and flowers. The expanding area architecture is the main difference between the different methods of hydroponic growth discussed in this book. Hydroponic recirculating systems, including those mentioned in this book, have that region that drains back into the reservoir. The reuse of irrigation water in hydroponics will significantly reduce the amount of water needed to grow a crop relative to the use of water used in conventional methods.

THE CROP

Plants that are nurtured in hydroponic systems will grow rapidly and yields more. Hydroponics removes the need for herbicides, and when combined with indoor growing methods can minimize or eliminate the need for pesticides. Hydroponic crops are also cleaner with reduced sprays, and no dirt than produce is grown using conventional methods. Most people know that water usage during the growing process can be minimized by hydroponics, but it is less widely understood that certain produce, such as lettuce, also need more water to wash than the total water needed to grow crops.

THE LIGHTS

Hydroponics is a common indoor growing technique, as it is clean and very efficient. When gardeners want to grow indoors, they also want to maximize yield in their small growing area, and hydroponic growing techniques usually accomplish that aim. A grow light is the primary equipment required to grow indoors. Indoor lighting has many choices, and each solution has its advantages. An increasing light can stimulate a wide range of desirable plant features, including enhanced flavour, increased nutrient content, increased plant pigmentation, reduced or increased plant height, earlier or delayed flowering, and increased yield, depending on the light intensity, duration and colour. Near to all the systems in this book can be used indoors when combined with a suitable grow lamp.

THE GROWING MEDIUM

Soil gardening and hydroponic soilless gardening are not enemies; each has its strengths and weaknesses. Blindly claiming that one is better than the other might be appealing to those who are strongly invested in one or the other approach, but doing so avoids the fact that both methods are very different.

Typically, the fertilizers used for hydroponics vary significantly compared to those used by soil gardeners. Hydroponic fertilizers surely provides everything required for great plant growth, while fertilizers intended for soil use should concentrate on only a few of the major nutrients since it is believed that most of the other nutrients are already present in the soil. Hydroponic fertilizers can work in the soil, but soil-based fertilizers will not have all the nutrients needed, but the nutrients are typically obtained from sources that may foul the water in a hydroponic garden. Manure, for example, is widely used for soil-based gardening but is almost never used in hydroponics. If used in a hydroponic greenhouse, most sources of animal-derived fertilizer such as manure, blood meal, bone meal, fish meal, and feather meal can produce terrible odors. One of the key benefits of soil gardening is the ability to use such animal-derived fertilizers, which are usually meat industry by-products. Soil gardening offers a perfect opportunity to use such by-products instead of going straight to a landfill for a better reason (growing plants). Many hydroponic fertilizers, and fertilizers in general, are generated using mined minerals and energy-intensive technologies, such as the Haber-Bosch process, which transforms atmospheric nitrogen gas (N_2) into ammonia (NH_3). This ammonia is used for the manufacture of fertilizers such as urea ($CO(NH_2)_2$) and ammonium nitrate ($NH_4 NO_3$). Traditional farming depends heavily on mined and synthetic fertilizers. Half of the nitrogen fertilizer added to crops is estimated to come from natural sources. Such fertilizers produce crops that feed billions of people. The advantages and disadvantages of synthetic and natural fertilizers are extremely complex. It can seem, when concentrating on one attribute, that one source of fertilizer is much superior to another, but the entire issue is far more complicated. For instance, the production of synthetic fertilizers has a much carbon footprint. However, Synthetic fertilizers, are far more concentrated in comparison with natural fertilizers and can be mobilized much more conveniently. Synthetic fertilizers are so clean and pure which results in great benefits in hydroponics. The use of synthetic fertilizers helps some farms to never discharge wastewater, resulting in tremendous water savings compared to conventional soil cultivation. The rather episodic advantages and disadvantages prove to me that no one has it perfect yet. There's plenty of room to learn from other growing methods and pool their benefits to establish ever more efficient farming methods.

CHAPTER TWO

THE EQUIPMENTS NEEDED

THE EQUIPMENT THAT WILL BE REQUIRED FOR a hydroponic growing system solely depends, of course, on what kind of system you want to operate. Hydroponics typically requires, except for the most simple systems, a pump for recirculating the water and fertilizer mixture. The recirculating water is very necessary as it is through this movement, and in some cases, an airstone with tubing, that oxygen from the ambient air is transferred to the liquid and then right to the plants. Along with the tubing and joining connectors, these pumps are the cornerstone of the system and perhaps the most important equipment you'll purchase.

ON IRRIGATION

For all we care, Irrigation might just be a fancy word for watering, but when you are talking about a hydroponic growing system, saying what it means can get quiet tricky. Whether you think of irrigation as providing nourishment or providing an infrastructure, the equipment you need to create the irrigation function really boils down to a couple basic items: a pump (with or without a filter) to propel and circulate the water through the system, and a series of tubes to convey the liquid.

WATER PUMPS

The major factors to consider when selecting a water pump are delivery height, target flow rate, and output tube size. Most systems simply need a pump powerful enough to deliver water to a specific height. For example, a grower selecting a pump for a flood and drain system can primarily focus on whether that pump has a maximum delivery height greater than the distance from the pump outlet to flood tray. Some systems function greatly when water is supplied at a target flow rate. A couple of systems that focus on the level of target flow rates are mostly NFT and aeroponics. For these systems, it is necessary to put into consideration how delivery height will influence flow rate. For instance, a pump that delivers 600 gallons per hour (GPH) at 4 feet high only delivers 200 GPH at 10 feet high. The number of emitters will also have an influence on the flow rate. It is generally better, and I must add, advisable, to choose a pump that may be slightly overpowered than a pump that could be underpowered. It is possible to reduce flow using valves, but it is not possible to increase flow.

- AIR PUMPS

Air pumps are primarily used to aerate, but they can also be effective for keeping nutrients evenly mixed in a reservoir. Aerating the nutrient solution can increase the dissolved oxygen. Although plants produce oxygen, they also use oxygen to perform a variety of tasks. One of these tasks is moving water through a filtration process in the roots. If a plant does not have adequate oxygen around its roots, then the plant will begin to wilt because it can not carry out the assignment of transporting water through the filtration process and up to the leaves. Increasing oxygen, as we know, in the root zone often increases crop yield and improves plant health.

Adding a check valve between the air pump and the air stone is an inexpensive way to protect your system from a potentially expensive failure. In the event of a pump failure, generally due to a power outage, water may siphon out of the reservoir down through the ¼ "tubing to the air pump. This could eliminate the air pump and flood the environment where the pump is. Many large pumps have more than one outlet sizes. Small pumps, however, are very useful in DIY hydroponic gardens, but they may only have a single outlet size. This small pump only connects to 5/16" tubing.

- ## THE VENTURI

A venturi attachment is a plain method to put air in a hydroponic system without the use of an air pump and air stones. A venturi could be attached directly to a pump or be installed inline in a section of tubing. Venturis take advantage of a phenomenon called the Venturi effect, which occurs when a liquid or gas flowing through a pipe moves through a constricted section, resulting in increased velocity and decreased static pressure. The venturi pump attachments have an intake tube positioned in the area of lower pressure. The decreased pressure creates a suction, which is used to pull air into the pipe. A pump with a venturi attachment can be placed on a reservoir wall to both circulate and aerate the nutrient solution. A 100' roll of ½ "black vinyl tubing A sample of ¼" clear vinyl tubing Air pumps are considered by airflow calculated in liters per minute (L / min). The target liters per minute for each hydroponic system is based on many factors, including reservoir size, water temperature, crop, and crop lifespan. In my experience, 1 L / min per 5 gallons is generally sufficient for most applications.

- ## AIR STONES

Air pumps deliver air through air stones, which come in a variety of shapes and sizes. Air stone preferences vary greatly by grower. I personally prefer flexible air stones and round air stones with bottom suctions. There are other ways to aerate a nutrient solution besides air pumps with air stones or water pumps with venturi attachments. Cascades or waterfalls are often the sole methods of putting air in nutrient solutions in NFT systems. Other more advanced ways this could be done include ozone generation and liquid oxygen injections.

•.TUBING

Not all irrigation pipes are alike. For most hydroponic applications conventional irrigation tubing used in landscaping is often very rigid and difficult to use. Black vinyl tubing is usually the preferred option for hydroponic irrigation, since it is lightweight, solid and easily connects to the preferred fittings used in hydroponic gardens. Of black vinyl tubing, the most common sizes are 1/4, 5/16, 1/2, 3/4 and 1 inch. Clear tubing for the irrigation lines is not recommended. Where the nutrient solution is exposed to light, there is always potential for algae production. Clear tubing could be a great spot for algae and, once algae grows, it is difficult to clean. Clear tubing is most common in aquariums as it is almost invisible and is more appealing to the esthetics. When aesthetics are not a big concern, it will perform as well as 1/4-inch transparent tubing.

FITTINGS

Flood and drain fittings make it possible for DIY gardeners to build their own flood trays from household materials such as plastic totes. Such fittings typically come in a package that includes a 1/2-inch fill fitting, a 3⁄4-inch drain fitting, extensions and two FITTINGS screens.

• **DIY gardeners:** may make their own flood trays from household materials such as plastic storage totes. Such fittings usually come in a package that includes a 1⁄2-inch filling, a 3⁄4-inch drain fitting, extensions and two panel fitting.

• **GROMMETS:** are one of the most popular DIY Hydroponic irrigation fittings. Grommets form a watertight seal around the fittings for irrigation. They can turn PVC tanks, plastic totes, buckets and more into reservoirs or hydroponic growing areas. Commonly available in 1⁄2 or 3⁄4 inches respectively. Tubing connectors work and look very much like the plumbing connectors you are used to using someone with experience doing home plumbing (except, of course, that they are much smaller).

POTS AND TRAYS
• **Net pots can** be square or round, and range from 2 to 10 inches long in general. This book focuses on uses of 2-and3-inch net pots, the most commonly used net pot sizes in hydroponic DIY systems.

• **Circular plastic** containers are typically the simplest to locate.

• **Square plastic pots** can help maximize space in a hydroponic garden by removing any gaps between plants. In rising trays, square pots are a common choice because they can be packed in tightly.

- **Growing bags** have long been used in commercial farms and start finding their way into home gardens. They may be hard to reuse, but they are certainly one of the cheapest pot choices. You can roll down the sidewalls of rising bags to change the pot volume. While when empty, the bag can look square, it fills out as a cylinder.

- **Fabric pots** are perfect for hydroponics because they drain easily, but they do not have wide holes that would allow the substrate to be removed. These are ideal for flood and drain systems as the water can easily soak in the soil and then drain rapidly. Simple to reuse fabric pots, too! Only empty the substratum, turn the bag inside out, let it dry, and brush any remaining debris away. We can also be put into a deep clean washing machine.

- **Terracotta pots** in hydroponics are not widely used, but this does not mean that they can not be used. Terracotta pots used in gardens are porous, enabling the passage of air and water through the walls, characteristics similar to a pot of cloth. Terracotta is heavy and fragile, unlike a pot made from cloth.

SUBSTRATES AND GROWING MEDIA

Hydroponic gardeners have a choice between quick growth high risk and slow growth low risk. The decision is based primarily on substrate porosity and the roots ' ability to breathe. Overwatering is one of the most common mistakes new gardeners make. An excess of water will drown the plant in heavy soil or in a poor-draining bowl. Although plants produce oxygen, they need oxygen too. The roots, in particular, require oxygen to accomplish a crucial step in water and nutrient uptake. The plant can not take up water without oxygen in the root zone, and the top Hydroponic gardeners will pick substrates that retain very little water to increase the oxygen available to the roots, but this involves regular or continuous irrigation. Some gardeners prefer using a substrate that holds more water to reduce the number of irrigation cycles needed. A substratum keeping more water provides some protection from power outages, pump failures and other possible causes of irrigation delays. When grown in a dry, sunny setting, a plant grown in a very porous substratum such as clay pellets will be weakened or die after a few hours with no irrigation. The same plant grown in coconut coir, a substrate which holds much more water, may be able to go without irrigation for a few days. The put-off for this development in safety is typically slightly slower development.

SUBSTRATES FOR STARTING SEED

This book focuses on bonded stone wool and polymer plugs made of peat moose and coconut. There are several other options for starting substrates, but these are two of the most beginner-friendly options as they have a decent water-holding ability and are hard to overwater.

• STONE WOOL

Commonly known in the United States as rock wool, stone wool is made by melting basaltic rocks and spinning the rock lava into fibres. Compared to the cotton candy but far less sweet. Disclaimer: Don't eat stone wool! Stone wool is one of the hydroponic substrates that is most common in both commercial and hobby hydroponics. It has a nice water retention and porosity balance which makes it perfect for new hydroponic gardeners, who often tend to overwater plants. Some substrates are not very forgiving for overwatering, but in general, stone wool will still work when overwatered— it may not have the best growth, but it will not typically destroy the crop. Stone wool comes in bricks, slabs, and loose.

• COCONUT COIR

Also known as' coco' coir, coconut coir is a-substratum made from coconut husks. It is a common substratum for traditional as well as organic hydroponic growers. If coconut is not washed properly during processing, it can have high salt levels which can damage salt-sensitive crops. Before using any coconut in a hydroponic garden it is a good idea to extract any residual salts and wash out any tannins that may stain the reservoir or growing area. The plant starts wilting. Having a plant wilting while sitting in water is really contraintuitive. Excess water may also make root disease more likely.

• COCO PEAT

A very fine coconut, also called a coconut pith or coconut dust, coconut peat can hold much water. It is also used as a supplement to or blended with peat moss. Unlike peat moss, Coco peat has a starting pH that is suitable to most vegetables without the need to add lime. Coco peat is frequently combined with perlite or another porous substratum, like peat moss, to lighten the mixture and increase drainage.

• COCONUT CHIPS

A chunky coconut, also called coconut croutons, has a strong water retention and drainage balance. These coconut chips can be utilized as a standalone substrate or incorporated in a combination. Using as a standalone substratum, coco chips can need regular irrigation, similar to growing in expanded clay pellets.

• PERLITE

Perlite is made of volcanic rock heating up like popcorn appearing. This expanded rock is so unheavy and has many business applications, primarily in the construction industry. Perlite is common, and used in horticultural sector because it is inexpensive, renewable, lightweight and perfect for heavy substrate aeration such as coconut and peat. It ranges from very fine to chunky in several sizes and can be used as a standalone hydroponic substratum.

• PEAT

Also referred to as sphagnum peat moss or sphagnum peat, peat is a partly decayed plant material extracted from bogs. It has the potential to carry a lot of water and yet when dry, it is lightweight, ideal for shipping. In general, peat has a very low pH of about 4. It is also combined with lime to elevate the pH for vegetables to a more appropriate level. Peat may be used as a standalone substratum but is used most frequently in a perlite blend. It is highly unavailable in North America, as mining this non-renewable resource in most parts of the world is severely restricted.

- **EXPANDED CLAY PELLETS**

Often called Hydroton after one of the original makers, and also known as LECA (which stands for light expanded clay aggregate), expanded clay pellets are pH neutral, inert, and one of the most common substrates for both hydroponic and aquaponic media beds. The pores in the pellets can hold some water, but overwatering clay pellets is difficult because they drain quite quickly. Often rinse clay pellets in a hydroponic greenhouse, before using them.

RE-UTILIZING SUBSTRATES

River rocks and clay pellets could be washed and reused, but other substrates in a hydroponic garden are typically difficult to reuse. In order]to boost water retention and drainage, most hydroponic gardeners can mix used coconut, peat, and perlite into their compost or directly into a conventional soil garden. In addition, some hydroponic gardeners will break up their used stone wool cubes and slabs into small pieces to blend into their typical soil garden.

EQUIPMENTS NEEDED FOR GROWING INDOORS

While there is no need for hydroponic gardeners to be indoors, they are usually associated with indoor growth. Growing indoors can sound better because there are less unforeseeable events, such as bad weather and pests, but indoor gardeners find a whole new set of challenges. A lack of adequate ventilation, poor temperature regulation, poor humidity regulation and inadequate light are some of the most common mistakes for inexperienced indoor growers. It is necessary to have the proper equipment to maintain a good indoor garden.

GROW TENTS

Grow tents have an enclosed space for monitoring, lighting and growing systems. Often having the proper growing indoor environment may be challenging, or the ideal growing environment may not be the same climate you want to have in the rest of your indoor space. Plants may like humidity ratios of about 50 to 80 percent, but people still tend to offer below that range in humidity. Growing tents in an indoor setting are a perfect way to separate the plants. In addition to maintaining a separate environment from the rest of the indoor space, a growing tent may maintain the bright light needed to grow the plants. Running rising lights for 20 hours or more a day is often helpful, but I can imagine people living in a tiny studio apartment would not be too happy to have a bright light on for 20 hours a day while trying to sleep. Growing tents may also allow gardeners to contain their pest management strategies to protect the crop, whether it is spraying or releasing beneficial predator insects. Grow tents are great for renters who can't change a growing space. Due to my enthusiasm at creating a growing space, I lost a few security deposits over the years without realizing that all the changes I made to the space might not make the landlord very happy.

TEMPERATURE CONTROL

A gardener can control the interior temperature with inline fans, depending on the climate outside a growing tent. Inline fans can be mounted inside the grow tent or outside it. Both of those configurations have advantages. An exhaust fan inside a growing tent is ideal for trapping crop odours, as it ensures that any air entering the growing tent passes through a carbon filter that removes all odours. This configuration is often referred to as a creating space for negative energy. Air passively flows from ducting ports through the room as the exhaust ventilator pushes out air.

• INTAKE FANS

An intake ventilator mounted on the outside will save valuable space in the grow tent. Air is forced into the grow tent in this system, and the exhaust passively escapes from the ducting ports. This high pressure growing room is perfect for pest control as it is difficult for pests to get into the growing tent due to the exhaust air. Occasionally, a negative pressure-increasing tent will suck in pests near any openings, but a positive pressure-growing tent can generate an external airflow that makes it impossible for pests to reach the increasing tent from somewhere other than the intake fan. There are several heavy-duty air intake filters that can prevent insects, bacteria, fungi and pollen from entering a growing space, such as the HEPA filter shown at left. Note: Lights can produce a lot of heat, and with ventilation fans, it can be difficult to control the heat. To indoor gardeners using very powerful lighting, using multiple lighting, growing in warm climates or growing temperature sensitive crops, air-conditioning units dedicated solely to the growing room are often required.

AIRFLOW

Inadequate airflow is one of the most common mistakes in indoor gardener beginners make. Luckily, it's one of the simplest ways to do. Insufficient airflow can lead to spindly, lanky plants, weak roots, tip burning, and increased likelihood of fungal problems in the crop (i.e., powdery mildew). A simple trick to test whether an increasing room has adequate airflow is to take a close look at the leaves to see whether they are moving visibly. Visibly moving leaves are a sign that there should be adequate airflow at that site, but the potential for "dead air" spots in a rising room still exists. Oscillating fans can help to reduce those dead air spots ' capacity.

GROW LIGHTS

One may date back to the 1800s the use of artificial light to grow plants. Growing lights were not always a practical choice, but advancements in lighting technology have been made in the last few decades that have made the use of growing lights available to amateur gardeners with gardens of any scale. There are many lighting options, but not all are well suited for your particular growing area; please check the many options before buying a growing light to prevent a potentially expensive error. These are potentially the most novice-friendly rising lights. They're also common and fairly cheap compared to other rising lights.

They consume minimal electricity and are available in many spectrums so that a wide variety of crops can be grown. They may not be suitable for crops such as peppers, which need intense light. Since they emit only small amounts of heat, they can be put in close proximity to the crop within a few inches, making them ideal for seedlings and young plants.

• HIGH PRESSURE SODIUM (HPS)

These are among the cheapest high-intensity lighting options available. HPS lights can produce plenty of heat which is fine in cold climates but hard to handle indoors without adequate ventilation and/or air conditioning. These are mostly used indoors for flowering crops and are useful in greenhouses to provide additional light. They are normally located just a few feet above a crop.

• METAL HALIDE (MH) AND CERAMIC METAL HALIDE (CMH)

MH and CMH are high-intensity lighting solutions often used in vegetative stages but often able to cultivate flowering crops. Light from MH bulbs appears blue, and many gardeners find working under them fun. The dominant blue light is also good for promoting compact growth. Rather of conventional MH bulbs, most rising light manufacturers focus production on the newer, more powerful CMH bulbs.

• LIGHT EMITTING DIODES (LEDS)

LEDs are very efficient and produce a lot of light using limited electricity. Relative to their light performance, they produce very little heat and are available in several different configurations, some suitable for mounting high above the crop and some suitable for placing very close to the crop. LEDs come in several different colours which can have a major effect on plant production. The white LEDs are less effective but more fun to work under than the red and blue LEDs, which casts a purple light that is perfect for growing plants but some growers find it aesthetically disappointing.

• ADDITIONAL LIGHT CHOICES

Other choices include, but are not limited to, induction lights, plasma lights and lasers, as well as many other lighting technologies. Some of these new lighting choices can be very costly, and may not be well suited for the hydroponic gardener who starts. However, lighting technology is advancing rapidly, and many of these options could soon be normal, just as LED lighting is moving rapidly to the fore of conventional HPS, MH and fluorescent lighting options.

LIGHTING ACCESSORIES
HANGERS

Lights can be attached or mounted directly to a crossbeam or ceiling with rope, wire, or chain. Rope ratchets are very popular with gardeners indoors because they make lights up and down very easy to move around.

RISE ROOM GLASSES

Many gardeners find it difficult to work under the orange HPS light, or the purple LED lights. Glasses with tinted lenses specifically built for such light sources are a perfect way to make working with those that lights more fun.

HYDROPONICS

PEST-MANAGEMENT PRODUCTS AND EQUIPMENT

Hydroponics can have an effect on pest pressure, but climate is the bigger factor on pest pressure. Hydroponic systems are mostly used in controlled settings such as indoor or greenhouses. Growing in a controlled setting gives the gardener the ability to remove pests from the crop entirely, but this can be very difficult to achieve. There are usually some bugs that get into the garden, and they can rapidly multiply once they get in. A managed garden with climate is perfect for plants and pests alike. If a bug enters an indoor greenhouse, it is in a good environmental setting with no predators. Pretty much the heaven of plague. There are many strategies for pest control, but prevention is always the best protection. Most of the pest management approaches can be used in a controlled setting or outdoors.

PREVENTIVE METHODS

Preventive methods include pest-exclusion strategies such as rooms for positive pressure growth and filters for HEPA intake, mentioned earlier in the section on Indoor Growing Equipment. Another custom of exclusion is to wear clean clothes before entering a growing room indoors to prevent bringing out pests from outside. Preventive approaches can involve choosing plant varieties that are ideal for growing climate and have resistance to disease, and providing those plants the water and nutrients they need to be healthy enough to resist disease.

• PHYSICAL

When prevention methods do not deter pests, and a disease is found in the field, physical pest management techniques are a safe, non-toxic approach for pest control. My favourite form of physical pest control is using a vacuum to eliminate any bugs that I find. Extra physical pest control strategies include the removal of entire plants and the use of sticky traps. They also use sticky traps to track pest rates.

• BIOLOGICAL

Pest management includes the use of predators, parasites and pathogens for protection of the populations of pests. One of gardeners ' most common biological pest-management techniques is the release of ladybugs. Biological pest control can not eliminate a pest population entirely but it can typically keep the pest population in check.

• ORGANIC PESTICIDES

Organic pesticides are usually considered less harmful than traditional/synthetic pesticides, but should still be used with caution. Often check the pesticide mark, including organic ones, to see if there is any approved personal protective equipment such as gloves, goggles or a respirator. Most farms use only organic pesticides to control pests to the max.

- **Conventional**, or synthetic, pesticides are rarely needed by domestic gardeners. Also, commercial farms that are not certified organic use organic pesticides very often, since they are very safe. Many of the traditional pesticides available to gardeners are as safe when used properly as organic pesticides.

PEST-MANAGEMENT TECHNIQUES

This is by no means an exhaustive list of pest control techniques, only a selection of my favorite pest management strategies in my backyard.

- **Vacuum**: This is a pesticide-free insecticide removal process.

- **Sticky Traps**: Sticky yellow traps are commonly used for trapping and tracking of aphids, whiteflies and fungal gnats. Commonly, blue sticky traps are used to capture and track thrips.

Beneficial insects

It can be difficult to control pests effectively with natural predators. There are several advantageous choices for insects; the following are a few of the most widely used pests in hydroponic homes. Growing room environment and presence of spray residues can impact beneficial insect effectiveness.

- **Lacewing (Chrysoperla carnea):** used mainly for the control of aphids, but may also be useful for whiteflies and thrips.

- **Coccinella septempunctata** (Ladybug): used to handle the aphids.

- **Mantis praying (Tenodera sinensis):** Consumes a large range of insects, including aphids. • Neoseiulus cucumeris (Predatory mite): used to monitor thrips and spider mites.

- **Swirski mite** (Amblyseius swirskii): For thrip power.

ESSENTIAL OILS

It can be very effective to destroy or repel pests such as mites, thrips and aphids. A few of the essential oils most widely used are garlic, clove, basil, thyme, rosemary and cinnamon.

Neem Oil

A neem tree-derived organic pesticide, this oil can repel insects and potentially kill them if added directly to the plague.

Azadirachtin

An extract made from Neem seeds which concentrates one of Neem oil's most potent insecticidal compounds. Azadirachtin repels Neem oil-like insects but, in many species, it also disrupts the molting cycle. Azadirachtin prevents pests from reaching maturity and reproducing in their juvenile stages.

Organic Pyrethrins

An organic chrysanthemum-derived pesticide. Another of the most effective organic pesticides, when administered at a high concentration, it is capable of destroying most insects quickly. Pyrethrins also may theoretically kill beneficial insects.

Bacillus thuringiensis (Bt): A beneficial microbe primarily used for caterpillar management.

Bacillus thuringiensis: Israelensis (Bti) subspecies A Bt subspecies that can have some biological regulation of fungal gnats.

Soap: Insecticidal soaps can be very effective in controlling whiteflies and aphids, even dish soaps.

Spinosad: An organic pesticide of the Saccharopolyspora spinosa bacterium. Good for handling caterpillars and thrips.

Streptomyces: lydicus A beneficial microbe that is protective against foliar fungi and root rot.

Potassium: Bicarbonate A very effective organic fungicide capable of breaking down powdery mildew problems easily. In hydroponic systems, can also be used to increase pH.

Sodium bicarbonate (baking soda) Very similar in effectiveness to potassium bicarbonate against powdery mildew; Plants can tolerate some sodium, but when exposed to large levels, they can exhibit salt toxicity or deficiency symptoms. Most gardeners may use sodium bicarbonate to combat powdery mildew and other foliar fungi in an efficient manner.

METERS

Many hydroponic systems employ a number of meters to track and help control the increasing environment. The meters calculate rates such as concentration and balance of nutrients, equilibrium pH, temperature and strength of light. Some function automatically, while others require that the hydroponic gardener create and maintain a routine monitoring system.

- **EC meters** of electrical conductivity is used to measure the concentration of fertilizers in a nutrient solution. EC meters are not essential to hydroponic growth but are certainly one of the most valuable devices. They're available from many companies in many shapes and in many price ranges. There are some really low-cost solutions available that I have seen growers work and have been keeping up for years. Personally, with my equipment, I'm not always the gentlest, and I want a powerful meter that can withstand violence. A truncheon EC meter is my go-to option at the moment because it doesn't need calibration, is waterproof and can manage misuse.
- **pH** While pH meters are not essential to hydroponic production, they are ideal to help hydroponic gardeners understand the condition of their nutrient solution. This is

also important to consider the pH of the nutrient solution when attempting to treat possible nutrient deficiencies. However, its meters are a little more temperamental than EC meters and should be handled carefully and well maintained. Otherwise, they may easily become unreliable otherwise simply break down. Always read the instructions on its probe to make sure you calibrate it correctly and do the routine maintenance needed to keep the probe accurate. There's a lot of variety on the market between its probes, and not all of them are equivalent. I've checked a lot of its meters, and my favourite currently is the Bluelab pH pen. An indicator solution can also be used to check it. Such indicator solutions also come as part of its monitoring package that involves solutions involving its up and down. A solution with its indicator can give an approximat, but it will never be as accurate as its meter. Many new hydroponic growers are starting with its control kit, with its indicator solution, because it is an inexpensive choice that can do the job.

LIGHT INTENSITY

It is extremely difficult, if not impossible, to determine the light intensity. There are several meters available to help gardeners track their light levels to assess if their particular crop is appropriate, adequate or too intense. Lux Meter Lux meters are usually the most reliable meter for light intensity measurement but not the most suitable meter.

Lux meters measure light on a scale which is similar to how the human eye perceives light. The human eye is most sensitive to yellow and green, while the plants are most sensitive to red and blue. Many of the crop light level guidelines are not based on lux; rather, they use photosynthetic photon flux density (PPFD), which is determined by meters of photosynthetically active radiation (PAR).

PAR Meter PAR is an acronym for radiation used in photosynthesis. PAR light comes within a spectrum of wavelengths visible to plants and can be used by plants to fuel photosynthesis. PPFD is an acronym for the flux density of photosynthetic photons. PPFD calculates how many photosynthetically active photons, expressed in μmol, land per second (s) in a square meter (m2); the unit being used is μmol / m2/s. PAR meters are the chosen meter in a horticultural setting for calculating light intensity, but they tend to be more costly than lux metres.

Daily Light Integral (DLI) Measurement A PPFD shows light intensity per square meter per second. A DLI calculation indicates the strength of light transmitted per square meter daily. DLI is a composite of all PPFD readings during the day, for every second. The standard used for this is mol / m2/d. DLI uses no μmol as the number would be enormous: 1 mol is 1,000,000 µmol. DLI is useful because it tests the light to which a plant has exposure all day long, not just at a particular moment. Indoors, measuring the DLI with a single PPFD calculation is relatively straightforward, since the light levels do not fluctuate as they do outdoors during the day. A PPFD reading of 100 µmol / m2/s indoors, for example, is translated to DLI with the following steps:

1 Multiply PPFD by 60 seconds to get total µmol per m2 per minute. Example: 100 µmol / m2/s= 6000 µmol /m2/minute

2 For µmol per m2 per hour, multiply this number by 60 minutes. Example: 6000 µmol /m2/minute= 360,000 µmol /m2/hour= 60 minutes= 360,000 µmol /m2/hour

3 Subtract this amount by the number of hours the lights are on. Example: 360,000 µmol / m2/hour= 7,200,000 µmol / m2/day

4 Finally, to convert µmol to mol, divide by 1,000,000. Example: 7,200,000/1,000,000= 7.2 mol / m2/day

Outdoor measurements can be made using a DLI meter. A DLI meter is designed to produce a DLI reading in mol / m2/day to complete the PPFD measurements during the day.

TEMPERATURE AND HUMIDITY CONTROL EQUIPMENT

A basic aquarium thermometer is often enough for temperature monitoring in a hydroponic reservoir. In most hydroponic crops, the target water temperature is 65 ° to 70 ° F, but it is definitely possible to grow safe crops beyond this range. Even most pH and EC meters measure the temperature of the water. Water temperature affects the EC and pH readings, so these meters must have a water temperature factor before accurate reading is provided.

A hygrometer thermometer which records high and low points daily is great for monitoring conditions in a greenhouse or grow space. Gardeners can spend a lot of time with their plants, but they can't be there all the time; a high and low point monitoring thermometer/hygrometer helps gardeners to adapt to day or night temperatures that they may not see while they're in the garden.

HYDROPONICS

CHAPTER THREE

HYDROPONIC GROWING SYSTEMS

DIY HYDROPONIC SYSTEMS ARE A great way to build a custom garden that is catered for your venue, crop and aesthetics you like. Most beginning hydroponic growers prefer to create their own systems because of the cost of retail systems, but I have found from personal experience that building DIY systems may not always be the cheapest choice, particularly if system design errors occur. I love designing original systems that are designed for unique locations, but designing original systems can also entail several costly errors. I bought things that didn't match or didn't hold after I put them in place, split, or gave enough light, or didn't give enough drainage. Basically, I learned a lot from my mistakes, and for that, I'm grateful, but I have spent a lot of time studying and making those mistakes. My experience, and my mistakes, draw from the following programs to save you time and money.

HOW TO CHOOSE A SYSTEM

Choosing which hydroponic system to install in your home requires that you take into consideration several variables and determine which one matters most to you. These include crop range, preferred hydroponic garden venue, maintenance requirements, ease of use and the amount of maintenance and maintenance each takes. Of course, initial costs are also significant, as are the costs of consuming electricity, inputs and other ongoing maintenance expenses.

• **Selecting a system by crop**

The first consideration when selecting a hydroponic system should put into consideration is what you want to grow. There are systems that can grow a wide range of crops (i.e., flood and drain), and there are some systems that function better with different growth patterns for crops. In this chapter, one of the first systems is a hydroponic bottle garden. This method works well for green leafy crops such as lettuce and basil but is bad for larger crops such as tomatoes. Also, you should recognize the variety of crops that you want to grow. Would you like to grow crops with a wide variety of nutrient demands and ideal pH ranges? Often, the best solution is to have several systems. The best thing about growing plants is that usually, they are easy to replace! Learn from practice and play with new crops. I give several recommendations in this chapter and helpful information on crop selection in the appendix, but these recommendations are not meant to discourage you from experimenting with them. Many crops are to grow outside of their ideal range in conditions. Plants are much more compassionate than we credit them for. Don't be afraid to fail; plant more seeds than ever!

Choosing a Location System

Hydroponic lettuce growing systems exist in space! No matter where you are, there is potential for hydroponic plant production. I've even got a hydroponic greenhouse in my RV. I give suggestions regarding the location for each of the systems mentioned in this chapter. Most such devices can be adapted for indoors, outdoors, small or large spaces.

Choosing a Location System Hydroponic lettuce growing systems exist in space! No matter where you are, there is potential for hydroponic plant production. I've even got a hydroponic greenhouse in my RV. I give suggestions regarding location for each of the systems mentioned in this chapter. Most such devices can be adapted for indoors, outdoors, small or large spaces.

Selecting a maintenance system Requirements

The ratio of plants to water volume is usually the most significant consideration for estimating maintenance requirements. A system with a small reservoir and a lot of plants would need regular maintenance as the grower will need to add water and change the reservoir with fertilizer because the plants are increasingly growing the reservoir water level. Systems with a high plant-to-water ratio often tend to accumulate an imbalanced nutrient ratio and need regular full flushes of the system. A further factor affecting the maintenance demand is the selection of crops. Depending on the variety, crops such as tomatoes, peppers and cucumbers may need trellising and pruning. Some crops grow very quickly and need to be replaced regularly, such as microgreens, so they will take a lot of work as they need to be planted so harvested each week.

- **Choosing aa system by its level of difficulty**

While I will never discourage someone from starting with an advanced hydroponic system, I know that many gardeners want to be productive from the beginning. Hard-hitting devices will have a learning curve. I just love to read! Possibly you too. But you may also enjoy simplicity and use a hydroponic system which has minimal moving parts and few chances of failure. Bottle Hydroponics, Floating Rafts, and Wicking Beds are fantastic, start-up-friendly systems that don't need electricity. Even beginner friendly are Media Beds and Flood and Drain, but they do have some moving parts that need electricity. Nutrient Film Technique, Top Drip, Aeroponics, and Vertical Gardens are not terribly difficult, but for a first-time hydroponic gardener, they may not be the best choice. The complexity of using a program is a personal opinion, and you may find it easier to use some of the less experienced systems. The only way to find out is to make them all up! Some hydroponic work now focuses on these rotating systems, but horticulturalists are still experimenting with static noncirculating hydroponics. Dr Bernard Kratky of the University of Hawaii is one of the most outspoken supporters of noncirculating hydroponics. He has done so much to promote the non-circulating hydroponics growth that his name has become synonymous with the technique. The Kratky style.

CROPS

The Kratky system has succeeded in growing a large variety of crops, from leafy greens such as lettuce to flowering crops such as tomatoes and potatoes. Many hydroponic gardeners prefer using the Kratky method to grow leafy greens and herbs because the larger crops can struggle with insufficient levels of oxygen in their root zone. The need for oxygen from the root zone for crops such as lettuce is much less than that for tomatoes. The crops ideally suited for bottle hydroponics remain short or grow upright to minimize the potential for the device to get too heavy and fall over. Basil, kale, Swiss chard, and lettuce are my favourites in bottle hydroponics, but with cilantro, dill, and other herbs I have also had success.

LOCATIONS

Outdoor, indoor, or greenhouse use of the Kratky process. In areas with heavy rainfall, it may be difficult to use a Kratky-style garden outdoors, since the nutrient solution may be diluted or washed away easily. Gardens in the Kratky style are perfect for off-grid gardens which do not have electricity access. The proper bottle hydroponic locations are tighter. The black paint used in this construct may result in excessive accumulation of heat in the root region. When you want to use outdoor bottle hydroponics, you would want to use a light-colored paint for warm-climate areas. My favourite outdoor use of hydroponic bottle systems is with a wall mounted bottle holder on a porch. This leaves the bottles in a semi-shaded environment and fantastic looks. Indoors, bottle hydroponics can be positioned almost anywhere— a kitchen counter, a desk, a windowsill, or even a wall mounted in a corridor above which grows light. Access to light is the only limiting factor when installing a hydroponic bottle device indoors.

If you are enjoying this book don't forget to leave a review, it's very important for me but what it is very important, you could help other people to find my guide: more reviews suggest more quality.

Have you done it? Thanks, let's move to our guide!

HOW TO CREATE A BOTTLE HYDROPONIC GARDEN

This hydroponic bottle is this book's easiest hydroponic garden and a great first step towards hydroponics. I enjoy developing this program when I do school visits with children aged 8 to 18. There are so many ways to customize the bottle with different paints and decorations, so it's easy to make your own garden. You may wish to find a bottle with an opaque exterior to skip the painting process to simplify the assembly of this device.

BOTTLE PREPARATION

The most important decision in this construct is the selection of bottles. The bottle should have a short handle, so that the plug can easily reach the bottle's main body. Pick a big bottle if necessary. Big bottles keep their water level higher, allowing the roots the time to expand into a nutrient solution until evapotranspiration causes the water level to fall. The following steps are for clear bottles so if using a nontransparent bottle, please skip to the next section.

1 Remove all bottle labels.

2 Remove a tape strip to the edge. To build a view window for the roots, this will be added later. Fold

the bottom of the tape strip to make removal after painting easier.

3 My favourite method of painting bottles is to position them on a post, but I also managed to dip bottles into water. Be sure there are enough coats of paint inside the bottle that light does not penetrate.

4 Remove the strip of tape until it has dried up the paint.

5 Any chalk art is best performed at this stage until the bottle is filled with water.

PLUG SELECTION

Either select a plug that snugly fits into the bottle's base, or pick a bottle with a good opening for your plugs. A stone wool plug can be cut to fit a smaller bottle, but this may potentially harm the roots of the seedling.

6. The plug should be wide enough to stay securely in the bottle's sopening.

7. Growing more seedlings than required helps you to select only the best seedlings for your hydroponic bottle with greater choices.

NUTRIENT SOLUTION AND TRANSPLANTING

Use of a fertilizer intended for hydroponic gardens is critical. I used Flora Nova Grow in my garden, but there are several other choices. To learn more about hydroponic fertilizer choices, check out the Plant Nutrition chapter.

8. Mix fertilizer with water using the recommended levels provided in the bottle or bag of fertilizer. To make it easy to test if the fertilizer has completely dissolved, mix the water and fertilizer in a separate bottle. Extra nutrient solution, if kept in an airtight container in a dark, cool setting, can be saved for a few weeks.

9. Fill the bottle full with a nutrient solution. Once the seedling is mounted, there is potential for some runoff, but this is preferable to too little water.

10. If you don't intend on using a wicking strip, you can now transplant the seedling into the bottle. The bottom of the plug should be seated in nutrient solution; if necessary, add more nutrient solution to ensure that the plug is fully saturated. Be sure the container is full if you don't use a wicking strip, as the plug will require several days of exposure to the nutrient solution before it can expand deep into the nutrient solution. The plug should not be placed too deep inside the bottle's base. To refill the bottle, you'll need to remove the plug, so leave enough of the plug

outside the container to make it convenient to remove in future.

11. Test to see if first week's plug is intact. You will need to add more nutrient solution in the first few days, depending on crop selection and climate, to give your plant a chance to grow roots long enough to draw water from the tank. A wicking strip is not necessary, but in the first week, it will help to reduce the risk of your seedling drying out.

ADDED WICKING STRIP

A wicking strip is useful in high and thin bottles, or with slow-growing crops. Consider the following steps: use a clear bottle for demonstration purposes, but it is not recommended to use a clear bottle to grow a crop, as it will encourage growth of algae.

12. Cut the burlap or cloth into a strip that is long enough to touch the bottom and about as wide as the seedling plug (usually 1 "to 2" wide).

13. Loop the wicking strip over the opening of the bottle.

14. To keep the wicking stripe in place, use the seedling plug.

15. Leave enough stone wool to show when refilling the bottle with nutrient solution to make removal quick.

16. A funnel could allow the bottle to be refilled without removing the stone wool plug completely. This may help to reduce the potential for damaging roots when removing and to reinsert a plug with a root system established.

17 When not using a funnel, take the plug out of the bottle very carefully.

18. Load nutrient solution into the container. It is best for young plants with poorly formed roots to fill almost to the top of the container. For older plants with larger root systems, filling up to three-fourths full is ideal, so that the roots have access to an air and nutrient balance solution.

After refilling, carefully reinsert the plug back into the tube. Verify that the roots are submerged in the nutrient solution.

MAINTENANCE

The majority of crops suitable to hydroponic bottles are rapidly growing and do not need much maintenance during their growth period. Long-term crops with several harvests, such as basil, can be grown as long as the container is kept more than half full with nutrient solution. To prevent nutrient imbalances in the solution, it is a good idea to clean out the container and refill it with fresh nutrient solution every month.

Next to chalk painting, I like decorating my hydroponic bottles with name tags and burlap scarfs. Covering the bottle's neck with a scarf can help hide any possible growth of algae on the seedling plug's surface. To secure burlap on the bottle neck, I use a hot glue gun.

Lighting Gardens with hydroponic bottle are ideally suited for indoors. They can be put on a windowsill, receiving natural light, or under a rising sun. Hydroponic bottles are a perfect addition to a work desk, under a tiny grow lamp.

TROUBLESHOOTING
WILTING PLANTS

• Test water level and add additional nutrient solution if the water level is low.

• The temperature of the water or of the air can be too high.

• Seek to apply wicking strip if the roots don't hit a nutrient solution.

PLUG IS DROPPING INTO BOTTLE

• Seek to wrap plug into cloth or burlap to build a snugger that fits into the bottle neck.

• Place the plug above the bottle opening so more stone wool is visible.

PLANT IS GROWING SLOWLY OR POORLY

• Hydroponic bottle garden may not be ideal for selecting crops.

• The plant will not obtain adequate sun.

• Use of a hydroponic fertilizer

FLOATING RAFTS

Floating raft hydroponics is a hydroponic Deep Water (DWC) subtype. The plant is kept at a fixed height for most conventional DWC schemes, and the nutrient solution is refilled to maintain contact with the roots. Floating hydroponic rafting helps the plant to keep in contact with the nutrient solution even as the water level decreases. Floating raft systems require very little maintenance and labour. When growing leafy greens, it's normal not to perform any maintenance on the method, not even adding water, from transplant to harvest.

CROPS

Floating raft hydroponics was used for large flowering crops such as tomatoes but is better suited for shorter crops with lower oxygen requirements in their root zone. With these larger flowering crops, conventional DWC systems are perfect as they provide space for the roots to reach air, and also use air pumps to heavily aerate the nutrient solution. Thousands of crops have been tested in floating rafts, and I am amazed at the flexibility of this growing process. On the next page, the sidebar lists some crops which can be grown in floating rafts.

LOCATIONS

Indoor, outdoor or greenhouse floating raft gardens can be built. We can have issues outdoors if not covered against storms. The rainwater dilutes the nutrient solution and washes the nutrients out. Floating raft systems also contain plenty of water, so this might not be ideal inside. There may be potential for leakage and indoor flooding if the device is not properly installed or constructed. Water is also very heavy, so floating raft systems with weight restrictions should not be mounted on floors. Aeration is of interest to floating raft systems. But that is not important for most crops. In 90 ° F water, I have grown beautiful lettuce and basil heads in floating raft gardens with no aeration. Such crops can benefit from aeration, both with faster growth and a reduced propensity for root diseases and nutrient problems, but without electricity floating raft gardens will flourish. If you want to keep your floating raft garden off-grid and enjoy the benefits of aeration, there are affordable options for solar-powered air pumps!

SIZING

Floating raft systems can be built for large fields or countertops. Very small floating rafts have the potential to become unstable when it comes to supporting big, top-heavy crops but are perfect for leafy greens. Large rafts may carry more weight, but they should be handled with caution when carrying heavy mature crops because when raised out of the reservoir, they can break under the weight. Some rafts are constructed of 2 pieces of 4-foot foam boards or 4 pieces of 8-foot foam boards. Many floating raft gardens are thus rectangular, with increments of 2 feet in width and increments of 4 feet in length. But don't feel restricted to rectangles; these foam boards can be cut in any shape. I've seen circular kiddie pools made into floating raft gardens with cut-to-size foam boards.

WICKING BED

WICKING BED GARDENS ARE VERY flexible and adaptable for a range of substrates, fertilizers and crops. Similar to the previous hydroponic gardens in this chapter, no electricity is needed from the wicking bed garden. The interface is extremely simple. Wicking beds take advantage of capillary motion, a natural phenomenon in which water will flow upwards toward gravity using friction and adhesion on the surface. A typical example is a wicking water upward from a cup of paper towel. The "cup" is the frame of a raised bed garden in a wicking bed garden, and the "paper towel" is a fine-textured substrate such as coco, peat, or dirt. A wicking bed frame is covered with a waterproof sheet, around 6 mil. Plastic painter, to prevent leakage and to stop the wooden frame from rotting. The bottom of the bed is lined with a quick-drying substrate such as clay pellets, river rock, or gravel washed. The bed bottom retains water or a wicked nutrient solution up to the fine-textured layer above. A barrier to the surface such as burlap or cloth stops the substratum from falling into the water reservoir. The inlet pipe makes it possible to fill the reservoir, and the overflow pipe avoids overwatering.

CROPS

crops that are resistant to wet conditions are perfect. This device might not be ideal for cactus. Wicking beds with many layers of different-textured substrates can be built to establish drier conditions while retaining adequate root moisture, but it can require some tinkering to work out the best combination for your particular climate, crop variety and garden size. It is often the size of a wicking bed garden which restricts crop selection. A wicking bed garden such as the one mentioned in the step-by-step guide might grow a large flowering crop such as a tomato or cucumber, but the garden's limited size would likely restrict it to just one.

LOCATIONS

Wicking bed gardens are usually used in greenhouses or outdoors. A wicking bed garden may be used indoors without making a massive mess by adding a collection bottle to catch overflow water or by directing overflow to a sink drain. The following design does not channel the overflow into a jar, and would not be appropriate indoors unless changed.

WICKING SYSTEM VARIATIONS

The wicking bed style is very flexible, and can be used in both conventional and hydroponic gardens. The design can be changed in the step-by-step guide to using conventional potting mixes and fertilizers that would not be suitable in other hydroponic garden designs. Below are some optional changes you can make to the configuration of the wicking bed to make it your own.

OPTIONAL MODIFICATIONS

• PVC can be used to make the inlet and overflow pipe instead of vinyl tubing.

• The frame may be a metal trough or plastic tote, rather than a wood liner.

• Instead of painter's plastic a pond liner may be used.

• Instead of using decorative wood the exterior may be finished.

• For larger crops, a timber trellis may be placed on.

• It is possible to mount a raised crossbeam above the growing bed to accommodate a growing light

NUTRIENT FILM TECHNIQUE (NFT)

Nutrient film technique (NFT) is a circulating hydroponic growing style that irrigates plants with a shallow nutrient solution stream in growing channels. NFT is one of the most common techniques for leafy greens which grow commercially. The ability to grow multiple plants on a small reservoir is one of the main advantages. NFT is very popular with rooftop growers because they can cover the entire roof in NFT channels using a small reservoir which does not exceed the roof's load-bearing capacity. Weighing 8.34 pounds, a gallon of water. Which means weighing over 240 gallons per ton! The weight of the water will add up quickly. Many home gardeners may also worry about heavy reservoirs, especially indoors. NFT is a very common hydroponic DIY technique because it can be configured in so many different ways. I saw NFT channels arranged in wall cascading patterns, in A-frame pyramids, and spiraling coils. Many DIY NFT systems are more effective than others— it can be simple to let innovative design take over and forget the fundamentals that make a good NFT garden. I encourage everyone to experiment but first read about NFT gardens ' possible weaknesses and complexities so you can avoid costly mistakes. Your NFT garden will be productive depending on crop selection, growing climate, channel length, channel slope, channel shape and flow rate.

CROPS

The most commonly used NFT crops are leafy greens, vegetables, and strawberries. Some crops have a good root system at maturity, but usually not enough roots to restrict flow in the NFT path. While growing bigger crops like tomatoes, peppers, and cucumbers, roots clogging the channels can be a concern. To accommodate the roots of these larger crops, some DIY gardeners use long PVC pipes (4 inches or more) or very big gutters. Feel free to experiment, but NFT is not, in general, the perfect method for growing large crops.

LOCATIONS

The capacity to irrigate several channels, without the weight of hundreds of gallons, in a small reservoir makes NFT common for indoor gardens. NFT is an excellent choice for rooftops, offices, balconies and flats. In general, NFT gardens have a pleasant flat canopy which is ideal for rising lights. Growing plants of different heights under a growing light is often difficult, as some that receive a lot of light while blocking the light for other crops, but this is rarely an issue with NFT indoor gardens.

NFT CHANNELS

The channels are constructed from 2-inch PVC pipe with 2-inch net pots in this design. Other common DIY options include 3-inch PVC pipe, rain gutters, and posts on vinyl fences. If the gutters are used, it is best to build a gutter cover to prevent the growth of algae in the canal. Flatbottom channels such as gutters and fence posts often funnel water to the channel sides, rather than down the centre. The separation of the water to the sides makes good communication between the seedling and the irrigation stream difficult. Gutters with grooves at the bottom often alleviate this issue by uniformly distributing the stream along the bottom of the pipe.

One very significant factor is the duration of the pipe. Most of the commercial NFT channels are between 4 and 15 feet wide. Longer channels often have wavering problems and need to be assisted at many stages. A sagging channel produces areas of stagnant water flow which can lead to a reduction in the oxygen available to the roots, an increase in the temperature of the water and an increase in the possibility of root diseases. For warm climates, long channels are not recommended, since they often have heat buildup issues. Once returning to the reservoir, the water must spend a long time in a long channel, and this prolonged channel time contributes to higher temperatures in the nutrient solution.

Gardeners in warm climates will concentrate on channels 8 feet and shorter unless the nutrient solution is cooled using a water chiller or another process. Often important is the slope of an NFT channel to reduce heat accumulation in the nutrient solution and to prevent stagnation of the nutrient solution inside the channels. A slope of 1 to 4 percent is acceptable; the slope used in commercial systems is usually 2 to 3 per cent. The device designed in this chapter aims for a 1-inch drop over a 4-foot (48-inch) channel to create a slope of 2 per cent.

FLOW RATE

Most NFT gardens target a channel flow rate of 1/2 to 1 litre per minute. With flow levels up to 2 1/2 litres per channel per minute, I have seen improvements in plant production. Attach the irrigation line to that channel and pass it to a measuring cup to calculate the flow rate per pipe. Either measure exactly how much water flows in one minute from that line, or figure out how long it takes to fill one litre and use that amount to calculate the flow rate per minute. In the Equipment chapter, the Irrigation section explains the method for calculating the minimum pump output to meet the flow rate requirements in a hydroponic garden; But because this detail is so significant, I repeat it here.

The key considerations to be considered when choosing a water pump are delivery height, target flow rate and tube size of output. Many systems simply need a pump that is strong enough to carry water up to a specific height. A grower choosing a pump for a flood and drain network, for example, should concentrate primarily on whether that pump has a maximum delivery height greater than the distance from the pump outlet to the flood drain. Those systems perform better when delivering water at a target flow rate. NFT and aeroponics are a couple of systems which rely on target flow rates. It's important to remember how distribution height would affect flow rate for these systems. A 4-feet-high pump delivering 600 gallons per hour (GPH) only delivers 200 GPH at 10 feet-high. Flow rate can also be influenced by the number of emitters.

It's usually safer to pick a slightly overpowered pump than a pump that may be underpowered. Using valves, flow can be decreased, but flow can't be increased. Example: A target flow rate for an NFT device is 15 GPH per path. The network is composed of 20 channels. It means that the pump must be capable of supplying 15 GPH to 20 channels, meaning 15 GPH for a total of 300 GPH for 20 channels. The channels are additionally 2 feet above the pump outlet.

TOP DRIP SYSTEM

TOP DRIP IS A HYDROPONIC technique that involves a wide variety of garden designs, all of which have one common feature: Irrigation lines carry water to the top of the substrate. The irrigation lines are often connected to flow rate regulators that produce a slow drip, and therefore top drip. Dutch buckets are one of the more common top-drip variants. Dutch buckets are pots with one drainage site closed-bottom. This drainage site is slightly raised from the bucket's bottom so that it can be set up to drain into a storage pipe that guides the nutrient solution used back into the reservoir to be recirculated.

• Appropriate locations: indoor, outdoor or greenhouse

• Size: medium to large

• Growing media: perlite or clay pellets

• Electric: Required

• Crops: leafy greens and large flowering crops, including tomatoes, cucumbers and peppers

Top drip systems provide nutrient solutions to the top surface of the substratum.

CROPS

Netherlands buckets are commonly used for large flowering crops such as hops, tomatoes, peppers, cucumbers, and eggplants. Some of those large crops can be cultivated in a Dutch bucket for a year or more. In Dutch buckets, leafy greens and herbs may be grown, but most hydroponic gardeners tend to take full advantage of their buckets by growing large flowering crops.

LOCATIONS

Bucket gardens in the Netherlands are usually outdoor or in greenhouses because the crops can get big. Most gardeners who use Dutch buckets install a trellis system next to the buckets so that plant growth can be guided upward and handled spatially. Indoor use of Dutch buckets is an option, but growth needs to be handled in a way that allows efficient use of that lights. Many rising lights may be placed vertically to light up a trellised crop. Most indoor gardeners set up a horizontal trellis and horizontally weave the plant growth to create a canopy even in height. For growing lights, a nice level canopy is perfect as it provides minimal shade of other plants and maximizes the use of light.

MEDIA BEDS

Media beds are a hydroponic garden design that is very basic. A growing bed is regularly flooded and drained from a reservoir using nutrient solution which is usually put directly under the growing bed. This arrangement is very similar to the flood and drain garden design discussed in the next section, the main difference being substratum placement. Media bed gardens simply load the substratum into the rising bed, removing pot requirement.

PROS

- Easy to grow a wide variety of crop

- Great for aquaponics, offers a lot of surface area for beneficial bacteria

CONS

- Limited to only a few substrates for filling the growing bed, hard to use fine-textured substrates

- Difficult to clean

CROPS

Media beds are perfect for long-term crops. Once extracting a plant from a media bed, it is very hard to extract the roots entirely. Many of these roots also break off, and this can accumulate quickly in a media bed while using fast-growing crops such as lettuce. Cutting and regrowing herbs are great choices because they can be harvested without killing the plant and destroying the root system. The media bed in the guide below is too small for flowering crops such as tomatoes and cucumbers, but some aquaponics media beds are much larger and can accommodate large flowering crops easily.

MEDIA BETTING

LOCATIONS can be planned for any venue. The media bed in the guide below is ideal for indoors but can also be put outdoors or in a greenhouse. Media beds put outdoors will have some issues if there is lots of rain— the reservoir will flood and the nutrients can be washed away— but the reservoir can be easily changed with fertilizer to restore the EC to a target.

SUBSTRATE OPTIONS

Expanded clay particles, expanded limestone, river stone, lava rock, aquarium gravel, and drainage gravel are only a few of the media-bed substrates. Be sure to use pH-neutral substrate made from large particles (evitate limestone). Every material used in a media bed is always prewashed. A rather coarse coir (coco croutons) can be used, but it's not ideal. Coco contains more water than conventional media bed substrates, so it is possible that the irrigation level will need to be lowered. Coco may collect more roots from the harvest plants, and more frequent cleaning may be needed. Coco also decomposes, and it will ultimately need to be replaced entirely.

IRRIGATION METHODS

Use fill and drain fittings, the traditional method for irrigating a media bed is. All fittings are secured at the base of the grow pad. The filling is flush, or almost flush, with the grown bed at the bottom and the drain fitting is raised just slightly below the grow bed level. The water reaches the grow bed after an irrigation process through the fill fitting, and the nutrient solution flows out into the reservoir through the drain fitting. The drain fitting prevents overflowing on the rising surface. Once the irrigation process ends, the nutrient solution drains from the media bed by flowing through the fill-fitting back into the reservoir. There are a few other common ways to irrigate a media bed, including bell siphons and U-siphons, but I would suggest sticking to the standard fill and drain fittings for beginners.

DRAIN AND FLOOD.

To capture the hydro juice, a flood and drain system needs a timer, a pump and a drain tank. The trousers run from the drain tank bottom to the pump inlet. The trousers run from the pump outlet to the hole at the top of the root chamber (high) flood end. The inlet of the pump is the drain tank under the bottom. As the drain tank fills through the pump, hydro juice flows through the pump inlet and up the flood hose with the hydro juice in the tank. It is to prime the pump because the pump is unable to drain air, it will force out just what flows in the inlet. The timer operates the pump for 1 minute, and about half of the root chamber is filled with hydro water. If the size of drain holes increases by chamber overflow. When a hose is used at the drain end, it must not allow hydro juice at the drain end to stand on. A recycle bin is suitable for drain tank (see the end of the section on Drip Feed to connect hose to drain tank). Putting the pump on the floor and the drain tank on bricks would lift the pressure preeminently enough.

EBB AND FLOW (FLOOD AND DRAIN)

The Ebb and Flow system operates by flooding the rising tray momentarily with a nutrient solution and then draining the solution back into the reservoir. Normally, this operation is performed with a submerged pump which is attached to a timer. When the timer clicks, the pump is pumped into the rising tray on a nutrient solution. As the timer turns off the pump, the nutrient solution flows back into the tank. Depending on the size and variety of plants, temperature and humidity, and the variety of growing medium used, the timer is programmed to turn on multiple times a day. The Ebb and Flow method is flexible and can be used for a variety of through media. Grow Rocks, gravel, or granular Rockwool will fill the entire grow tray. Many people prefer to use individual pots filled with growing medium, which makes it easier to move or even move plants inside or out of the network. The key drawback of this kind of device is that there is a susceptibility to power outages as well as pump and timer failures with certain forms of growing medium (Gravel, Growrocks, Perlite). The roots will dry out quickly on interruption of the watering cycles. Using through media that retains more water (Rockwool, Vermiculite, coconut fiber or a good soilless mix like Pro-mix or Faffard's) will somewhat relieve this problem.

AEROPONICS

Aeroponics is a hydroponic process which is really exciting. It offers very fast growth potential and large yields while using very little water. Within aeroponics, there are two main categories: high pressure and low pressure.

High Pressure

The Construction Guide below illustrates how to build an aeroponic high pressure greenhouse. Most hydroponic farmers, when they hear the word aeroponics, think of designs with high pressure. A pump, mostly PVC, is attached to a main irrigation pipe, and misters are inserted into the PVC tube. In the PVC pipe, the pump creates pressure which helps to generate a fine mist.

- **High-pressure:** aeroponic designs for rooting cuttings or "clones" are very common. Fine nutrient solution mist provides a perfect environment for new root growth.

- **Low Pressure:** Aeroponic gardens should not hire misters. The aeroponic "water" is often produced by moving through perforated disks to the nutrient solution and/or causing splashes near the plant roots.

Aeronomic low-pressure systems typically have less moving parts, and are less likely to clog.

CROPS

Virtually, every crop is aeroponically growable. I saw papayas rising in airborne systems! Leafy greens and herbs are the most popular crops for aeroponic systems but do not feel limited to these choices. If larger flowering crops are growing, be sure to consider how to help the plant. Plants grown in pots can be assisted by securing their roots to the substratum (to some extent). Without a substratum, there is no physical support for the plant roots and a top-heavy plant may lean or fall over if it is not supplied with support, such as a vertical or horizontal trellis. Long-term crops are also more likely to experience a power outage or malfunction of equipment that could easily damage roots or destroy plants that could have taken several months of care.

AEROPONICS

LOCATIONS is ideal for any location. Aeroponic gardens can be small and fit on kitchen counters, or they can be huge vertical towers over 15 meters tall. DIY aeroponic gardens can often have leak problems, and they should be checked before they are put in a leak-sensitive location.

VERTICAL GARDENS

VERTICAL GARDENS Use soil and hydroponic growing methods in all shapes and sizes. Vertical gardens are popular with gardeners with limited space, because in a given footprint, they can optimize the available area of growth. Vertical gardens are also common as installments of living art. Going to a bar, restaurant, office, or school is becoming increasingly popular and seeing a vertical garden being used as an edible art installation.

When selecting a vertical garden, there are a few things to bear in mind. Firstly, not all crops are well suited to this method of production. Big, highly heavy crops such as tomatoes, eggplant and peppers will not have the help they need if they are grown in a vertical garden. Some of the hydroponic vertical systems are ideally suited for leafy greens, spices, and strawberries. The second major factor is the light requirement of the crop chosen. Vertical gardens when incorrectly built or placed are infamous for having light problems. Vertical structures often cast shadow on crops below. Lower crop light may not be a problem during the summer when there is a lot of sun, but this can be a problem in lower light conditions.

Whereas this book focuses on hydroponics, when choosing a vertical garden design, hydroponics is not the only choice. It would be simple to change the garden shown in the following project to use a potting mix and receive hand waterings. Personally, I think that putting in the initial effort to create a hydroponic system pays off in the long run as I don't have to remember watering my plants, just for each one of them-that's DIY! Below are some of the common hydroponic horticultural setups.

AEROPONIC TOWERS

Low or high-pressure Aeroponic devices may be used. A vertical high-pressure aeroponic garden will typically have a main line of irrigation in the center of a wide tube or square. The main irrigation line will have evenly spaced foggers or misters which emit a fine mist for the plant roots placed inside the outer tube or square. Such devices require a reasonable amount of pressure and may be clogging-prone. An irrigation system that uses misters or foggers requires the use of an unprecipitating, high-quality fertilizer. The grower also needs to be careful about leaves and roots dropping into the system, as these can break down and block emitters. In the centre of a wide tube or square, a low-pressure aeroponic vertical garden may also have a main irrigation but will only release the nutrient solution at the top of the garden. The solution of nutrients then falls through a series of disks which disperse the water. Tower Garden is a very common aeroponic vertical low pressure device. DIY variants of this system are possible but actually purchasing a full system is often beneficial.

DRIP TOWERS

Drip towers come in different shapes and sizes too. Nearly all of them consist of either a vertical post or a bag full of an inert substratum such as perlite, coco, or stone. The ZipGrow tower is a vertical drip tower which has gained much attention in recent years. This uses a plastic matrix in the center of a square panel, and a capillary pad.

FLOOD AND DRAIN GROW RACKS

A growing vertical device in commercial farms is the flood and drain grow racks. Below is a picture of Growtainer's vertical flood and drain network. Many farmers are developing their own implementations of those systems. Most of these are made from metal storage shelves, flood tables, and lamps. The inclusion of shut-off valves for each level is critical when designing your own flood and drain grow rack. These valves will allow you to change the flow to each stage, so that they fill in approximately the same time. Also important is the height between levels and the location of the lights. Most of these rising racks are between rates of 18 to 24 inches. I recommend that you use fluorescent T5 or LED lighting bars. With rising racks, the most common problems I see are inadequate light and poor airflow. Spindly, stretchy method is one of the main markers of low light levels. Firstly, it can be a hassle to drive the Ferris wheel with a motor so the plants can be dipped into a nutrient solution. Second, gravity and water weight are perfect on a Ferris wheel for plants to pass around. Second, using the fast-draining pans. Stone wool and/or perlite are usually fine choices for such structures.

NFT A-FRAME

The NFT A-frame structure is composed of NFT channels arranged in A shape. They have both benefits and disadvantages. The cons are the potential to increase plant sites within a given footprint. The drawbacks are an irregular light distribution and potential difficulties in flow rate. When you intend to construct an A-frame NFT system, follow the same slope and flow rate guidelines, as shown in the NFT project. Alternatively, using 1/4-inch shut-off valves to regulate flow between all rates for each pipe. The use of 1/4 inch shut-off valves for the rain gutter garden is further defined in the project.

RAIN GUTTER SYSTEMS

These are one of my favourite seedlings, and they're fun to create and customize.growth. The seedlings reach out to obtain more energy. Sometimes, spindly seedlings are best removed and start anew. Airflow may help strengthen seedlings too. A small clip-on fan will shake the seedlings gently, enabling them to grow stronger stems and roots that are better developed. For crops such as head lettuce, inadequate airflow can sometimes contribute to tip burning.

HOW TO BUILD A RAIN GUTTER GARDEN

This system is one of the most complex systems in this book, but that's because I concentrated on the final system's aesthetics. Personally, I want a device that looks so good that a garden visitor wouldn't think it's a DIY project right away. To make this device easier to install, you can skip the paint job, use vinyl tubing to attach troughs, and reduce the number of levels. Alternatively, one can think of this system as a model for a much larger system. I cut my channels to 33 inches long, but it could easily be modified to have 10-inch channels. This could be higher in other stages too. In addition, it is important to consider the pump size when adding more vertical rates. I prefer using shut-off valves to control the flow for each stage. Over-sized pumps also help to reduce debris potential by clogging the irrigation lines.

NEEDED MATERIAL

Frame

1   2 × 10" × 8' board

2   2 × 4" × 8' board

1 lb. #10 x 2½" exterior screws

20 gal. Reservoir

Troughs

1 10' vinyl rain gutter

6 White vinyl gutter hanger

3 White vinyl K-style end cap set

50 L Coarse perlite

Irrigation

3 ½" rubber grommet

3 ½" elbow connectors

10' ½" black vinyl tubing

6' 1½" PVC pipe

3 1½" PVC coupling

4' ¼" black vinyl tubing

1 ½" stopper

3 ¼" double-barbed connectors

3 ¼" shutoff valve

1 Submersible water pump (800 GPH)

3 Active Aqua screen fitting

The frame width must be greater than the reservoir so that the reservoir can comfortably fit between the vertical supports.

1. Transfer the 2 rollover 10 "rollover 8 ' board to the sawhorses and fasten with clamps. Measure and mark two 30" segments to be used as the frame foundation.

2. Draw square lines of cutting for each 2/10 section.

3. Wearing work gloves and eye protection, cut the 2 some 10"/ 8' board with the circular saw into two 30" pieces.

4. Measure and mark a 5' segment on both boards. Putting the boards on top of each other will help to ensure that they are cut to the exact same length.

6. Cut the two 2 / 4"/ 8' boards along the specified lines to produce two 5 ' segments and two 3 ' segments.

7. Measure, mark, and cut a 30 "segment from phase 6 using one of the 3'2" as well as 4 "segments.

8. Push the segment 30 "2x 4" over the sawhorses, fasten with clamps, and mark the middle.

9. Using the drill bit of the 2 "hole saw to create a 2" hole in the designated center of the segment 30 "2x 4."

10. Attach the two 5' 2= 4 "segments to one of the 30" 2= 10 "frames. Use two screws on each side. 11. Attach to the 30" 2= 4 "segment the other side of the two 5' 2= 4" segments. Using two at each side of the screws.

12. Cut two small 2x 4 "pieces from the remaining 2x 4" wood to create the support legs for the frame.

13. Cut the angle support legs.

14. Secure the support legs to the base using two screws. Secure the support legs to the vertical supports with one screw.

15. Set up the frame in a level location. Verify that the base and top crossbeam are level and square. 15. Secure the support legs to the vertical supports with one screw.

20 Measure and mark three 33 "segments in a 10' vinyl gutter.

21. Use a hacksaw and/or heavy dutyscissors to cut the three 33" segments.

22. To clear any burrs from the ends of gutters, use the deburring method.

23. Mount the 33 "gutter segments and fasten with clamps on sawhorses. Measure and mark 161/2" gutter middle.

24. Using a phase bit to make a 3/4 "hole in the marked gutter centre. Placing this hole is very necessary! It should be located closer to the curved edge of the gutter. The hole center is about 2" from the flat gutter back. Deburr the hole but make sure the hole doesn't widen too far.

25. Place the 1/2 "grommet into the hole. Use the deburring device to enlarge the opening, if the opening is too low.

26. Repeat steps 24–25 for each of the 33" gutter parts.

27. In each gutter segment, insert a 1/2 "elbow into the 1/2" grommet.

28 Cut out 3 4 "1/2" tube segments. Stick to the elbows.

Attaching Troughs to Frame The troughs are 18 "apart from the top beam in this design, starting 3." It gives under the lowest trough 18 "of room for the 11"-tall reservoir.

29. Label 3, "21" and 39 "from the top beam on each of the vertical 5' 2x 4" segments.

30.. Push the gutters into the hangers.

32. Remove the end caps.

IRRIGATION ASSEMBLY

The irrigation design of this system uses one main PVC pipe for both drainage and concealment of the tube used for water supply. It is possible to simplify the assembly of the irrigation by eliminating the PVC pipe and by using 1 pipe. Mark the approximate intersection between the 1 1/2 "PVC pipe and the 1/2" drainage tube that comes from each gutter.

34. Draw a second set of marks about 1/2 "above each gutter point. Distinguishing between the marks made in the previous step and this step would be necessary.

35. Remove the 1 1/2" PVC pipe and fasten it to the sawhorses using clamps. Drill a 5/8 "hole into each of the three points in step using the phase bit

36. Drill a hole into each of the three marks created in phase 34 using the 1/4 "drill bit. This will be used for the 1/4" water supply line.

37. At the bottom end of the 11/2 "PVC pipe (the end nearest to the 1/2" hole), drill four 13/8 "drain holes. This pipe should sit on the bottom of the reservoir and drainage water should flow through those holes into the reservoir. Use the deburring method to clean the edges. 38. The 11/2" PVC couplings should be mounted above the 1/4 "hole for the lower two levels and between the 1/4" and 5/8 "hole for the top level.

39. Cut the 11/2 "PVC pipe at the marks made in phase 38.40. Link the cut PVC segments with couplings.

41. Bring the installed PVC mainline back into the device with the top coming out of the guide hole in the 2 / 4" wooden crossbeam.

42. Test to see whether 1/2 "drainage tubes can be inserted into the main line of PVC from all three levels. If not, shorten segments in the main line of PVC until all 1/2" drainage tubes suit their respective holes in the main line of PVC. The PVC couplings may create gaps (sometimes large ones) which increase the pipe's total length, discarding previous measurements.

43. On the 11/2 "PVC pipe, mark the top of the 2x 4" wood crossbeam.

44. Cut the top section of the PVC pipe and cut along the mark made in phase 43. It stops the PVC pipe sticking high above the top wood crossbeam.

45. Change the mainline PVC to a flat service. Select 1/4 "vinyl tube, 1/2" vinyl tube, 1/2 "stopper, 1/4" double-barbed connectors, 1/4 "shut-off valves, irrigation line hole punch, and scissors. 46. Cut 3 10" segments and 3 2 "1/4" vinyl tube segments.

47. Place the 1/2 "vinyl tube next to the main line of PVC. It may be beneficial to use clamps to keep the line straight.

48. Insert the 1/2" stopper into the end of the 1/2 "vinyl tube at the top of the main line of PVC.

49. Start punching holes in the 1/2" vinyl tube with the punch of the irrigation line.

50. Insert 1/4 "double-barbed connectors into the three holes in the 1/2" vinyl tube created in phase

51 Attach the 2 "and 10" parts of 1/4 "tubing with the 1/4" shut-off valves.

52. Remove the PVC mainline from the 11/2 "couplings.

53. Insert the 1/4 "tubes mounted from phase 51 into 1/4" holes in the main PVC side. Insert the 10 "section into the PVC, so that the 1/4" shutdown valve stays outside the main line of PVC.

54. Attach the 10 "section of the 1/4" vinyl tube to the 1/4 "double-barbed connector in the 1/2" vinyl tube starting from the top of the PVC main line.

55. Slide the 1/2 "vinyl tube down the main line of PVC and proceed to connect the 10" segments of the 1/4 "vinyl tube to the 1/4" double-barbed connectors in the 1/2 "vinyl tube.

56. Reconnect the main line segments of PVC.

57. Slide the lower end of the 1/2" vinyl tube through one of the drain holes created in phase 37.

58. Place the assembled PVC main line back into the vertical system with the top held in place by the guide hole in the 2 × 4" wood crossbeam.

59. Insert the 1/2 "drain lines from each gutter into the corresponding 5/8" hole in the main line of PVC.

60. Attach the bottom end of the 1/2 "vinyl tube to a pump in the reservoir. 59. Insert the 1/2" drain lines from each gutter into the corresponding 5/8 "hole in the main line of PVC.

PLANTING

Test prior to planting of the irrigation system. Fill the reservoir with enough water to cover the pump, turn on the pump and check that water is obtained at each point. Adjust the flow to each stage with the shut-off valves set. The irrigation check will also allow the irrigation lines to be washed and will capture any loose plastic particles left over from the assembly. Dump the water to the check.

61. Place the Active Aqua screen fittings over 1/2 "drain grommets before filling each trough. 62. Pre-rinse the perlite in a bucket. It will help keep the device clean. Fill each trough with perlite. 63. The amount of space left at the top of each trough will depend on how many plants you intend to install.

• Disassemble the irrigation and reassemble it to that level. Shorting 1/4 "irrigation line can also be helpful.

CHAPTER FOUR

STARTING AND GROWING HEALTHY SEEDS

Vigorous SEEDLING or root cutting is always one of the greatest challenges for new hydroponic gardeners. The optimal conditions for germination or root establishment depend on crop selection. Rinse with a half-strength nutrient solution, and wash the stone fur.

2. Let any excess nutrient solution run through the mesh bottom tray away from the seedling pad.

3. Place the bottom mesh tray onto the bottom strong tray. The stone wool should be warm to the touch but not in sweat.

4. Place the solid base tray on the heat mat for the seedlings.

5. Seed that sheet.

6. One pelleted seed per plug should be sewn.

7. Basil also yields more when one uses three to eight seeds per plug. Even if not put directly in a dibbled hole, Basil will always germinate well.

8. Lettuce mixes that use raw seed (not pelleted) yield more and look better when they use three to five seeds per tube.

9. Where appropriate, plants such as tomatoes, peppers, cucumbers, and eggplant should be seeded for two per tube. Identify the smaller plant in the plug until the seedlings emerge, and extract it by pinching and pulling. Using two seeds per plug and removing one later in each plug increases the likelihood of successful seedlings. When there is no germination of one seed, then there is a backup.

10. know the types. Using plant markers or make a note on a sheet of paper; either way, keep track of what varieties you are planting.

11. Misting the seeds will help ensure good contact with the stone wool and have adequate moisture to germinate. Misting with pelleted seeds is very beneficial as they often fail to retain enough moisture simply by making contact with the stone wool.

12. The heat pad sensor is inserted into the heat plate. Weave the thermometer of the controller into one of the dome vents for the humidity and drop it into the stone fur.

13. Place the humidity dome on the tray and pull the thermometer cord to any excess slack.

14. Set the control heat mat to a minimum desired temperature. The appendix in the Crop Selection Charts lists different target germination temperatures.

15. There should be no need to contact the seedling tray during the first few days. The initial rinsing/soaking of stone wool should provide ample moisture for some days.

16. Remove the humidity dome once the seedlings have germinated by 50 percent. This will be after 3 to 5 days for most vegetable crops. Leaving the humidity dome on too long may increase the risks of fungal diseases and death from seedling.

17. Stone wool will feel heavy when wet, and it will be noticeably lighter when irrigation is required. By raising up the tray to gage the weight, it is best to gain a sense of how much water is in your seedling pad. Based on air temperature and crop age, irrigate with a nutrient solution when the tray feels light; sometimes, this is every 2 to 4 days indoors. The seedlings do not need to be irrigated at all, depending on the climate, so they may be ready to be transplanted into your hydroponic garden within 1 to 2 weeks before irrigation is needed.

ROOTING CUTTINGS IN STONE WOOL

There are several different ways of rooting cutting and the procedure among these methods varies greatly. This guide covers some of these variations; please try to see what works best for you, your crop, and the special cloning environment you have.

HYDROPONICS

1. Rinse with a half-strength nutrient solution, and wash the stone fur. Place it inside the solid base tray. Pick cuttings, use sharp pruners.

2. Shorten the cuttings to 4 "to 7," allowing an internode cut of 45 degrees below.

3. Wear gloves even when using rooting hormone.

4. Pour some of the rooting solutions into a separate jar to avoid the whole bottle being potentially polluted.

5. Dip the cutting end into the rooting hormone, and allow any excess rooting solution to drip off before transferring the cut to the cube.

6. The cutting can be done in many ways inside the cube:

   a. The traditional approach is to insert the cutting into the cube approximately 1 "deep through the pre-dibbled hole.
   b. Another alternative is to make a smaller dibble hole so that the cutting fits more snugly into the cavity. It is advantageous when using thin cuttings because it reduces the amount of contact between the stem and the stone wool. \
   c. I generally aim for 6" cuttings so I can have a couple of inches of stem submerged in the nutrient solution.

7. Assemble the cloning system and place under the grow light if using indoors.

8.	Fill with half-strength nutrient solution or use a hydroponic fertilizer specifically for rooting cuttings (sometimes called a "clone solution").

9.	Plug in the air pump and water pump.

10..	Use a soft collar to hold the cuttings in place. Make sure no leaves are stuck in the collar.

11.	Evenly space cuttings in the cloner and cover any unused holes with a collar.

12.	After 4 to 7 days, most cuttings show evidence of roots. Some plants root more slowly than others and may need to stay in the system longer.

13..	Plants with established roots are ready to be transplanted into a hydroponic garden. Simply remove the collar, and your new plant is ready to go.

TRANSPLANTING PLANTS STARTED IN SOIL

The option of using a soil-started plant in a hydroponic garden is often very attractive to new hydroponic gardeners because it makes it possible to purchase plants from a garden center or use plants from their existing soil garden. It is definitely possible to transplant soil-started plants into a hydroponic system, but it is not the best way to source plants for a hydroponic garden. The process of rinsing off the soil from a plant's roots usually involves some root loss and damage, which increases the potential of exposure to root diseases. If the hydroponic garden uses small irrigation lines (¼ inch or smaller), it is possible that any soil particles not rinsed off from the transplant may clog the irrigation lines. I would not recommend transplanting a soil plant into a hydroponic system if you are not okay with the possibility that the plant may not survive the process. Now that the disclaimers are out of the way, I must personally say that I really enjoy the process of washing off the soil from a plant's roots.

1. If possible, prune off all the fruit and some of the vegetation from the plant. Less fruit and vegetation means less need for water uptake and less demand on the root system. It is important to reduce the demand on the root system because it might be damaged in the rinsing process and unable to deliver the water and nutrients required for the full-size plant.

2. Pour off any loose soil from the top of the transplant.

3. Remove the plant from its pot.

4. Gently dunk the root system into a bucket of water.

5. Gently shake the plant to wash off soil from the roots.

6. Use your fingers to loosen up the roots to expose soil clumps trapped deep within.

7. It may be necessary to dump and refill the bucket multiple times to get all the soil off the roots. A watering wand with a gentle flow can help speed the process.

8. Pick out as much soil and debris as possible without ripping up the roots.

9. Clear some space for the transplant.

10. Insert the transplant and cover the root system.

11. Water in the new transplant to improve root contact with the substrate.

CHAPTER FIVE

PLANT NUTRITION

PLANT NUTRIENT UPTAKE

PLANTS CANNOT TELL THE DIFFERENCE between natural and synthetic fertilizers. Plants have specialized pathways that only allow them to uptake a very short list of ions and simple molecules. In traditional soil-based gardening, these ions and molecules are often derived from manure or decaying plant matter broken down by a series of biological processes. For example, nitrogen is primarily only available to plants when present as ammonium (NH_4^+) or nitrate (NO_3^-). In manure, nitrogen can be present in a wide variety of forms, including organic nitrogen (Org-N), ammonia (NH_3), ammonium (NH_4^+), hydrazine (N_2H_2), hydroxylamine (NH_2OH), nitrogen gas (N_2), nitrous oxide (N_2O), nitric oxide (NO), nitrous acid (HNO_2), nitrite (NO_2^-), nitrogen dioxide (NO_2), nitric acid (HNO_3), and nitrate (NO_3^-). Bacteria present in the soil can transform these forms of nitrogen into the specific plant-available forms of nitrogen. The process of breaking down a raw nutrient source like manure into simple molecules and ions available to the plant is dependent on many factors, including bacterial populations, soil temperature, and water content. In traditional hydroponic fertilizers, nitrogen is applied in its plant-available forms (ammonium and nitrate) and there is no need for bacteria to process the fertilizer into plant-available forms.

Plants grown in soil are constantly searching for nutrients. Their roots are on a scavenger hunt for nutrients spread through the soil. The roots generally find nutrients dissolved in water in the soil, often called the soil solution, which can then be picked up by the roots. The availability of nutrients in the soil is dependent on not only the presence of nutrients, but also the moisture in the soil, pH of the soil, distribution of nutrients in the soil, the cation exchange capacity of the soil, and more.

On the other hand, plants grown in hydroponic systems can have constant access to nutrients. The nutrients are evenly dissolved in water to create a nutrient solution, similar to a soil solution. Any time the plant needs water or nutrients, they're available. This allows a plant to reach its full potential without needing to expend energy searching for nutrients or being stunted by the inability to find essential nutrient.

FERTILIZERS

Fertilizers can be an extremely difficult issue. This is one of the most common issues that I tackle while dealing with commercial hydroponic growers. Decades ago, almost all farmers had to mix more than 10 ingredients together to create a hydroponic fertilizer solution that met all the nutrient requirements of their crops. That included a lot of chemistry, laboratory experiments and tablets! Many commercial growers still make custom fertilizer blends using several ingredients today but pre-mixed fertilizers are increasingly being used. Such pre-mixed fertilizer mixes allow growers to simply buy two or three different bags of fertilizer to produce a mixture that meets all the nutrient requirements of their crops. Hobby hydroponic fertilizer manufacturers have further streamlined the cycle by offering one-part fertilizer options. One-part fertilizers are as simple as concentrates for fruit punch. Using the rate on the fertilizer bag or bottle, simply apply the fertilizer powder or liquid concentrates to a given volume of water.

ORGANIC HYDROPONIC FERTILIZERS

Organic hydroponics is possible but for new growers I would not suggest it. Before venturing into organic hydroponics, it is important to have some experience with hydroponics and understand how plants can function under normal conditions. Aquaponics is one of the most beginner-friendly choices for a new hydroponic grower looking to grow organically. Aquaponics is a mixture of aquaculture and hydroponics, or fish farming. The fish waste in an aquaponic system is broken down into a series of biological processes to produce nutrients available to the plants.

You may be tempted to play with an organic fertilizer that was made for conventional gardening but this sometimes leads to foul-smelling mess. Most organic fertilizers are made of manure from cattle, or meat industry by-products. Those fertilizers in a hydroponic system will easily turn rancid. The nutrient solution begins to smell bad, and the machine gets coated in goop, causing the gardener to wash and clean the machine regularly. Most productive organic hydroponic fertilizers use plant-derived nutrients, such as sugarcane. I have developed many productive organic hydroponic systems using Pre-Empt, a molasses-based fertilizer.

FERTILIZER SOURCES
LIQUID OR DRY FERTILIZER

A variety of types exist within traditional fertilizers. Most hydroponic gardeners make the first choice between the liquid or dry fertilizers. Liquid fertilizers are also easier to use, as they are easy to calculate and need minimal mixing, but liquid fertilizers are often more costly than dry fertilizers. Many liquid hydroponic fertilizers are essentially a dry, water-mixed fertilizer and then sold in a container. Liquid fertilizers are also less dense than dry fertilizers, and are more costly due to higher shipping costs.

ONE-PART, TWO-PART, AND SOME PARTS

Some hydroponic fertilizer companies are trying to build a product line with tons of add-ons, but these add-ons are mostly unnecessary for healthy plant production. Many new hydroponic gardeners get carried away with modifications to the fertilizer and do more damage than benefit. Loving a plant to death is surprisingly easy. New hydroponic gardeners want to give every glamorous product they see to their plants but too much love will destroy your crop quickly.

One-part fertilizers such as those mentioned below will grow healthy crops 99 per cent of the time without any modifications. Most single-part fertilizers are developed either for vegetative production, such as lettuce and young plants, or for reproductive development, such as mature fruit-bearing tomato plants or any other flowering crops.

There are many fertilizers which come in two or three sections. Such multipart fertilizers are somewhat different from add-on goods, and they come in two or three sections because, when combined in high concentration, other nutrients appear to bind together. It is called precipitation binding. The typical culprits are phosphate calcium, or sulfate calcium. When these nutrients combine, a precipitate is produced which looks like sand. The sand would fall to the reservoir's bottom and become inaccessible to the crop. Most firms market their fertilizer in two parts: one with calcium (along with other nutrients) and the other with phosphorus and sulfur (along with other nutrients).

Often one-part liquid fertilizers have poor shelf-life because the nutrients start to produce precipitates that accumulate at the edge. Shaking a one-part liquid fertilizer bottle before buying is always a good idea to test if there is a solid chunk of fertilizer precipitate at the bottom.

The advantage of two-and three-part fertilizers is their ability to change the nutrient ratio. Many of the fertilizers that were produced for the hobby hydroponic gardener recommended ratios of each of the ingredients for different growth stages.

Big commercial hydroponic farmers and universities make other parts of the fertilizers. Generally, each of these additions comprises one or two of the thirteen basic plant-growth nutrients. These recipes for fertilizers frequently contain ten or more different ingredients. If you are really excited about stoichiometry and want to learn about advanced hydroponic fertilizers, then studying the Hoagland solution is a great place to continue. The Hoagland solution and the several variations of updated Hoagland solutions are based on the initial recipes of hydroponic nutrients produced in the 1930s at the University of California.

MEASURING FERTILIZER CONCENTRATION

The concentration of fertilizers in a hydroponic nutrient solution can be measured in many ways. The recommended unit of measurement can differ according to country and application.

ELECTRICAL CONDUCTIVITY

Electrical conductivity (EC) is a measure of the capacity of a material to bear electric current. The ability of water to conduct electricity is the reason why it is extremely risky to swim during a thunderstorm, or to use an electrical appliance in a lake. Surprisingly, pure distilled water that has no mineral content is a very bad conductor indeed. Pure distilled water is not common, and due to its mineral content nearly all water sources have a certain degree of conductivity. In hydroponics, with the introduction of fertilizers, growers increase the mineral content of the soil. These fertilizers improve the ability of the water to conduct electricity in a consistent pattern. For this reason, EC is a perfect way to estimate the concentration of fertilizer in a solution of hydroponic nutrients. EC is usually expressed per centimeter (mS/cm) in millisiemens. Some countries, mainly Australia and New Zealand, can use factor of conductivity (CF) instead of EC.

Parts per million (ppm) refers to the amount of an element in a given volume of water, usually expressed in milligrams per liter. Complete dissolved solids (TDS) meters typically equate with Ppm. The conversion chart has a few different ppm columns since ppm can be viewed in several ways, depending on the ppm meter/probe manufacturer. To new hydroponic growers attempting to reach a recommended ppm, this can be a great source of uncertainty as they may be unsure whether their meter is calculating ppm using the same definition as the recommended ppm.

I consider using an EC meter to prevent uncertainty. That being said, it's really curious that there are so many different definitions of ppm. As previously mentioned, EC tests how well a nutrient solution conducts electricity, and EC increases as fertilizer is applied to a solution, but not all fertilizers equally increase the EC. Some nutrients have little effect on EC, while others have an effect that is very important. For example, an EC reading of 1 mS/cm could mean that calcium is 400 ppm or that it could mean that phosphorus is 620 ppm.

Within the nutrient, solution nutrients are present as ions, and certain ions are stronger electricity conductors. Nearly all ppm meters calculate the EC of a solution, and then convert that number into ppm by multiplying the EC by a conversion factor provided by the manufacturer as a ppm approximation. This means that the supplier must anticipate which nutrients are to be used in the nutrient solution to decide how its meter will translate the original EC reading into ppm. Again, if given a choice between EC and ppm, please use an EC meter to avoid this ambiguity.

CHAPTER SIX

SYSTEM MAINTENANCE

EVEN THE most simple hydroponic gardens would require some upkeep at some point. Maintenance of hydroponic systems involves everything from tracking and modifying nutrient concentrations to daily flushing of the system, and even scrubbing pots and reservoirs periodically.

MANAGING THE NUTRIENT SOLUTION

A hydroponic nutrient solution can be handled from many directions. Choosing your hydroponic garden management strategy will rely on crop quality, tank size, garden design and personal preference. I always choose the choice that needs the least amount of time even though this may have a minor impact on the growth rate or crop quality, but you may want to control your nutrients more closely to maximize production. The following management strategies are grouped according to the commitment that they need.

LEAST EFFORT: SET AND FORGET

Install the reservoir on the fertilizer bag/bottle with the suggested fertilizer cost per gallon. Change the pH if it is, or are not, well beyond the target range. Enable the crop to grow until it's ready to harvest or until plants have too low water level to access the nutrient solution. This approach may work well in floating raft systems for leafy greens, and may work in other systems if they have a sufficiently large reservoir relative to the number of growing plants. I have produced a remarkable number of beautifully looked crops using this technique of minimal effort. If used for crops for long growth periods, such as tomatoes, peppers, cucumbers and other flowering crops, this method of management may have issues. Try the top off method if you want to use reduced effort and grow crops that have a longer period of time before maturity.

LITTLE EFFORT: TOP OFF

This approach is similar to set and forget, except the grower simply adds water to retain the original level as the water level decreases. Over time, this approach may dilute the concentration of nutrients in the reservoir, and there could be nutrient shortages on the crop. This method will work for fast-growing crops with low demands for nutrients, such as microgreens, leafy greens and some herbs. This method often works depending on the system for some larger crops, but there is some risk of overdiluting the nutrient solution, particularly when using a small reservoir.

SOME EFFORT: TOP OFF AND AMEND

The most common approach to maintain a nutrient solution in a hydroponic garden is to top off the reservoir as stated in the previous process, then add more fertilizer to the reservoir to hold a goal EC. See the appendix for growing hydroponic crops, for example, target ECs. The grower changes the pH of the nutrient solution using either an acid (pH down) or base (pH up), after applying fertilizer to meet the target EC. In some stores, there are many easy-to-use pH down and pH up items, and there are DIY alternatives that are often less ideal but certainly functional. Some hydroponic gardeners use vinegar or lemon juice for pH down, and others use baking soda for pH upwards.

You will need an EC meter, a hydroponic fertilizer, a measuring cup, a pH meter, pH down and pH up adjustments, and a pipette (eyedropper) to top off and amend the solution in your system.

FLUSHING

EC is a perfect general guide in a hydroponic reservoir for nutrient content, but sadly, it doesn't tell the whole story. Plants suck up not all of the nutrients at the same time. Many nutrients will accumulate over time, and others will rapidly deplete, resulting in an imbalanced nutrient solution. Large commercial hydroponic farms submit water samples to check facilities to get precise quantities of each nutrient in the reservoir, and the grower then changes the fertilizer inputs to that. To make these changes to the fertilizer requires complex chemistry and a thorough understanding of the unique nutrient requirements of a crop. The much simpler solution is to wash out a hydroponic device regularly. Flushing is the method of extracting the current nutrient solution and fresh water refilling the system, and then adding new fertilizer. The frequency of flushing depends on several factors, including the nature of the crop, climate, method, fertilizer, and water. Most gardeners succeed in finding out the flush frequency using the following thumb rule: "Flow a reservoir when the quantity of water added to the top of a reservoir is equal to the size of the reservoir." Example: A 40-gallon reservoir loses 5 gallons per day for evapotranspiration (plant transpiration and evaporation of the reservoir). To top off the tank for water loss the grower adds 5 gallons daily to the tank. The grower adds a total of 40 gallons after 8 days (8 days, 5 gallons= 40 gallons), the same amount of water as the original capacity of the reservoir. Every 8 days the grower will flush out the reservoir.

This rule of thumb is very conservative, and when using conventional hydroponic fertilizers many farmers will flush less frequently. Nonetheless, this rule is useful for having a general guideline. The water which has been flushed from a hydroponic system does not need to be drained down. The old nutrient solution is used by many gardeners to water their potted plants, raise beds, lawns or trees. A conventional garden is a perfect complement to a hydroponic garden and can be a habitat for old fertilizer solutions, composted seeds, and substrates.

CLEANING

Hydroponic growers may use their gardens with a variety of items. Dish soap is typically the safest and easiest alternative. Some additional choices available to hobby hydroponic growers include household bleach (use 1/2 to 1 ounce per gallon of water), isopropyl alcohol (70 per cent or higher) and hydrogen peroxide (3 per cent is usually sufficient; stronger concentrations are accessible but must be treated with caution, so read and observe product labels).

CHAPTER SEVEN

COMMON PROBLEMS AND TROUBLESHOOTING

NOW THAT YOU HAVE ALREADY decided to be a device designer, an indoor gardener and a maintenance worker, it's time to learn to be a doctor. Here's a brief guide on how to treat and overcome your rising hydroponic system.

NUTRIENT DEFICIENCIES

Typical nutrient deficiency and toxicity recognition guides display a single leaf with symptoms, but these can easily lead a gardener to over-correct an issue or incorrectly correct a problem. Nutrient toxicity or deficiency is most frequently due to nutrient solution/substrate pH, environmental conditions, crop age, or pathogen presence. Test to see if:

- All plants of the same variety exhibit similar symptoms before suggesting the issue is nutrient linked.

- The pH is not low (below 5.0) or high (above 6.5) in the target crop range;

- The EC is within the target crop range.

- The air temperature for the crop is within the target range.

- The temperature of the water is within the optimal crop range, not below 55 ° F or above 85 ° F;

- All crops are getting sufficient airflow. The leaves should pass clearly.

- Pest-free crop.

- The amount of light is within the target range.

- The nutrient solution is produced using a hydroponic fertilizer.

If the answer is yes to all of these conditions, then the nutrient-related issue is likely a battle you must now fight. Nutrient related issues can also be remedied by dumping the nutrient solution out and restarting the device.

CHLOROSIS AND NECROSIS

Chlorosis in plants is the degradation of chlorophyll, that is, the green pigment. Chlorosis may be used to characterize leaf yellowing from several causes including shortages in nutrients or damage to the pests. Necrosis means the death of plant tissue. Plant diseases or deficiencies in nutrients often begin with signs of chlorosis which lead to necrosis.

- **Early-development interveinal chlorosis**

Interveinal chlorosis on new growth also suggests a deficiency in iron or other micronutrients. Some hydroponic fertilizers provide plenty of iron. Therefore, iron is rarely the problem. Iron shortages are usually caused by too high a pH. Many crops are "iron-inefficient," and are struggling to capture iron. Basil is an example common to an iron-inefficient plant. If basil is grown in a high pH nutrient solution, often just over 6, it can exhibit interveinal chlorosis on new growth suggesting an iron deficiency. The leaves showing this form of interveinal chlorosis will not recover, but if the pH is changed and/or iron replaced with the nutrient solution, future growth will return to normal. \

- **Older leaf chlorosis**

Chlorosis on older leaves can result from a few different scenarios: nitrogen deficiency: nitrogen is a major component of chlorophyll, the green leaf pigment. Plants will take up the nitrogen from chlorophyll and transfer it as required across the field. If a nitrogen deficiency is identified by the plant, it will transfer the nitrogen in its older leaves to its new growth. Deficiencies in nitrogen can occur when the crops grow at low EC. Occasionally, modern aquaponic gardens have problems with nitrogen deficiencies.

- **Natural senescence:** Senescence is leaves dying due to old age which is normal. It is not unusual in mature plants to see some bottom leaves die from natural senescence. If both young and old plants are in the hydroponic greenhouse, test to see whether only the older plants exhibit chlorosis on older leaves; this would suggest natural senescence.
- **Magnesium deficiency:** This is similar to a nitrogen deficiency with older leaves showing chlorosis, but interveinal chlorosis with necrotic spots and/or necrotic edges of the leaves may be present. Most magnesium shortages can be remedied at a rate of 1/2 to 1 teaspoon per gallon with magnesium sulphate (Epsom salt).

TIP BURN

Tip burning is potentially a calcium deficiency, but it occurs more often even though calcium is present in the nutrient solution. Calcium is essential to plant-cell wall formation. Calcium absorption by the plant may also fail to keep up with the development of new cells when a plant grows rapidly in an atmosphere with intense light and warm conditions. There are several ways to solve this issue.

• **Another variety to seek.** Many varieties are highly sensitive to tip burning while others can grow well under established conditions.

• **Boost airflow** to boost transpiration and improve calcium absorption in the crop;

• **Using the less nitrogen fertilizer** to slow the production.

• **Giving less light to the crop** by adding shade or changing a rising light to slow production. • With calcium. It often helps, but the majority of hydroponic fertilizers do have enough calcium.

INFESTATIONS ALGAE

Growth of algae is typically not a concern, but it can lead to other problems. Algae can steal some nutrients from the nutrient solution, but this is not normally a big issue. The main concern is that algae can serve as a source of food for fungus gnats and shore flies. To monitor the growth of algae reduces sunlight access to the nutrient solution. Algae growing on the seedlings ' surface is also a sign of overwatering, but it is not typically an issue that will impact plant growth significantly.

FUNGUS GNATS AND SHORE FLIES

Fungus gnats feed on fungi, algae and tissue from plants. In general, adult fungal gnats do not pose a threat, but the larvae may damage crops. The larvae feed on plant roots, making the plant vulnerable to pathogens such as Pythium and Fusarium. Shore flies are very similar in appearance to fungus gnats, but their larvae do not feed on plant roots. Shore flies do not damage crops, but they can be irritating for sure. Fungus gnats have a body shape that resembles a mosquito with long legs. Shore flies resemble a fruit fly rather than a mosquito. There are several ways of managing fungal gnats and shore flies; the following are only a few strategies:

• Eliminate from developing area algae and rotting plant matter.

Make useful nematodes such as Steinernema feltiae.

• Use products that contain the bacterium Bacillus thuringiensis israelensis (Bti) for pest control.

• Use organic azadirachtin-containing pesticides

• Use organic pyrethrum / pyrethrin-containing pesticides

APHIDS

Aphids do not normally destroy plants, but can damage crops by distorting growth or spreading viruses. A sticky honeydew at the leaves is the most common sign of aphids. This honeydew can attract bees, or maybe a growth site for fungi. Insecticide soaps are effective in handling aphids. This is very useful for products containing azadirachtin or pyrethrum/pyrethrin.

THRIPS

Multiple thrips or a single thrips can exist. The term thrips is both the plural form and the singular form. Thrips harm typically occurs as spots on leaves, deformed growth of flowers and/or distorted growth of new leaves. Thrips can be a destructive insect. Having crops or crop types that are less appealing to thrips is also easier to find. A variety of biological pest control strategies, such as introducing predatory insects such as green lacewings, parasitic mites, parasite wasps, and minute pirate bugs, can be used. Organic spinosad-containing insecticides can be highly effective on thrips. Additional solutions include azadirachtin or pyrethrin-containing insecticides, or an insecticide soap.

SPIDER MITES

The two-spotted spider mite is the most common found in gardens. It is primarily a flowering crop epidemic, including tomatoes, aubergines, cucumbers, and strawberries. Early harm typically shows up on the top surface of the leaves as a dull speckled appearance. This can progress to leaf chlorosis and drop in the leaf. Visible webbing on leaves has poor infestations. Quite frequently, spider mites prey on a plant from the upper leaves. Spider mites prefer dry weather and are attracted to heavily fertilized crops. If used preventively, the predatory insects can be very effective. Commonly used are the predatory mites Phytoseiulus persimilis and Amblyseius fallacis. Neem oil and insecticidal soaps can also help manage mite populations. Often, do two applications around 5 to 7 days apart when using an insecticide on the mites. Insecticides can not regulate mites in the egg stage as well, so spacing out applications helps to eliminate the mites in full.

SEEDLING PROBLEMS

One of the most difficult phases in the process for new hydroponic gardeners can be growing a safe seedling. Here are only a few of the reasons you may encounter poor germination, seedling death or low quality seedling.

- The substratum is too damp and the young seedlings rot (common with fine coconut and heavy soil).

- The substratum is too dry.

- Sowing long, thin stems, due to low light.

- Other seeds have poor germination rates, obviously.

- Many seeds are pretty temperature sensitive.

WILTING

In many hydroponic systems, there is the possibility of water. Some crops benefit from having the root zone dry out between the irrigation cycles. There are many techniques to assess when a crop should be watered, including the finger test, lift tests, and meters. The finger test simply brings a finger through the soil surface to search for moisture. Finger tests are less effective on large containers, which can hold much deeper moisture than a finger would test. For large potted plants, a list test is more accurate. Only raise the pot to see if its water weight is high. Water is very hot, and when the pot is light and in need of water, it will be noticeable. There are a number of moisture meters that can also support but a finger test and/or lift check is often enough.

MUSHY BROWN ROOTS

The dark-mushy roots are the dead roots. The following are some potential causes of root death:

• Insufficient oxygen in the root zone due to heavy soil overwatering, no aeration and/or high water temperatures.

• Very high EC damage or a nutrient deficiency.

• Damage from extremely high or very low pH.

• Very warm water temperatures, often seen over 90 ° F in hydroponic systems.

• Damage from sanitizers used to disinfect structures that had not been properly rinsed prior to replanting.

• **Pathogenic root rot**. There are plenty of pathogens all clumped together under the same root rot general name.

Delete plants which have dead roots. Quite often, a system needs to be washed completely and sanitized to avoid the presence of pathogens. Until replanting, seek to correct environmental conditions that are suitable for root rot. Can help to increase the flow rate in NFT. Might help to improve aeration in a floating raft garden. This may also help to fill the tank, or to install a water chiller.

CHAPTER EIGHT

TROUBLESHOOTING

- PH related problems
- Pest
- Water
- Overuse of nutrients

- Distance to light-sources

- Root bound

- Algae

- Mold

- Fungus

For just the device itself, growing for hydroponics can be very difficult enough, but when problems occur with the plants, this may turn into a major problem. There, we'll go through some of the more common problems that can hold you back from rising, so that you'll be able to correct and avoid these pitfalls in the future.

PH RELATED PROBLEMS

When you feed and take care of your plants properly, and you find any signs of disease, then the problem is more than likely related to it. Below are some signs of its related symptoms: leaves burning even though nutrients are properly supplied, leaves twisting and stop developing all together, and yellow spot. Checking the water solution's PH with a PH meter will easily solve its problem, and then change it according to its adjuster available in almost any plant shop. Its associated conditions are among the most common and simplest to handle. Its tester is a must when growing with hydroponic systems and all hydroponic cultivators should also have at least one.

PEST

Plagues will surface! True–that in an indoor garden. It is uncommon to get pests unless the plant comes from outside. Anyway, the problem here isn't how to get pests but how to deal with them.

TABLE OF PESTS AND PREDATORS:

Type	Consequences	Latin name
White-fly	White-fly lays eggs on the underside of leaves leading to shiny sticky leaves	Parasitic wasps – Encarcia Formosa
Thrips	Small yellow speckles on leaves; because of the miniature size of thrips, they are often only observed after the damage	Amblyseius cucumeris and Orius laevigatus
Aphids	Responsible for Fruit contamination and virus carrier	Aphidius colemani and Aphidoletes aphidimyza
Aphids	Destroy plants by feeding on plant tissue	Phytoseiulus persimilis
Beet Armyworm	Elimintes tomatoes	Hyposoter exiguae

There are several different types of pests that can destroy your crop entirely, so it's important to know how to detect a pest problem and whether you have one and how to eradicate it. I suggested that every plant be tested very well at least 1-2 times a week. This is where pests typically thrive, particularly underneath the foliage. It's pretty easy to detect some sort of bug infestation, the typical signs are tiny holes on the leaves, broken bits of leaves, leaf-to-leaf webs and even the bugs themselves being seen. If carefully and frequently reviewed, you will spot any bugs on your plants and treat them accordingly before it becomes a problem. There are several ways of solving this issue.

Among the most successful are the purchase of an organic plant pest control spray, can be found in virtually every plant nursery, and growing shop... Another way is to spray your plants in a spray bottle with a 10-20 per cent neem oil/water mix and spray the whole plant down. Then a very good home remedy that works really well is to mix a dab of simple dish soap in a spray bottle, fill with water then spray thoroughly, making sure that the leaves underneath are sprayed properly. Return within 1-2 hours and rinse off with plain water.

You may also use insecticides and pesticides to control and address the issue of plague—this is not the ideal hough approach. First, all pesticides are poisonous (whatever the manufacturer claims) and, second, pests establish pesticide tolerance. How, then, are we doing?

We can introduce other insects which are our pest predators. This may sound nuts to you, but that's how the real world food chain works. This predators ' population depends on the availability of food; so as your pests dwindle in number, so do your predators do.

You can also use traps of insects (sticky papers); you can buy or make these. Cut out cardboard strips to make them, paint them with whatever colour you want; let them dry out completely. Now, add Vaselineor some jelly of petroleum; use these strips every 3 to 4 sq.

WATER

If plants are overwatered, they can rapidly begin to grow root rot, which can escalate out of control quickly before you will even know that there is a problem. It may result in a plant ceasing to grow new seeds. Those are the main symptoms of overwatering: the leaves and roots wilting can become brow and mushy and stick a little. Signs of rain underneath: leaves begin to drop rapidly and finally dry up. Simply offer less or more water to correct those problems. A bit of a tip to consider is that having less water than more is often better for a plant.

OVERUSE OF NUTRIENTS

This is perhaps the most common issue when using hydroponics, the symptoms include: leaf "fire," leaves cupping inwards, and occasionally a spotting of the copper colour may occur. If this happens just change your reservoir and use straight water for a couple of days, then re-apply nutrients and test the PH levels. By using hydroponic nutrients, you must be very careful by calculating the amount to be placed into your tank, follow the prescribed nutrient feeding schedule exactly, or less. All nutrients should have a manner of usage somewhere on the bottle.

DISTANCE TO SUN-SOURCES

The outer parts of your leaves closest to the sun will spot this common problem, turning a brown colour and gradually warping it. If your hand gets hot when you're putting it under the sun, then you know it's too close and get up. Today, on the other hand, the signs are very clear when plants are too far away from a light source or the light production is not enough. The plants are going to tent to "stretch" into the sun, basically stretching and elongating themselves. Lights being too far would also cause very slow growth and ultimately cause growth to stop entirely, and they will not recover if you would push your plants into sufficient light.

ROOT BOUND

This is spotted by slow growth, and roots from the top of your hydroponic system begin to come. This can be easily identified and corrected by simply making sure that you have enough room for roots in your hydroponic system and checking the roots regularly to make sure they have more than enough space. You'd be surprised how quickly they can develop and expand.

ALGAE

This is common among nearly all hydroponic growers and can be seen inside the tubing and within your tank. The cause of this is exposure to sunlight. To avoid this issue, try to shield with a reflective material as much as possible as your device and its irrigation tubing. Reflective tape performs very well in this regard. Another way to avoid the growth of algae is to clean your hydroponic tank too regularly and to tub it at least every few weeks. Doing so and preventing any light leaks should ensure that algae do not grow.

MOLD

Mold is typically a common problem when growing indoors, since the air is closed. Mold is typically caused by excess air moisture, and/or inadequate air circulation in your growing field. The signs are a white substance which builds on your growing medium and the plant itself. The first thing to do when you have mold is to check the humidity with a hydrometer, and test if the humidity levels are normal. First I'd test to see whether your ventilation fans are air circulation proper, it's always nice to have at least 1-2 fans for draining air out of the space, 1-2 fans pulling in fresh air, to 1-3 fans blowing directly on your plants. Make sure you follow all of these criteria and your mold issue will be a thing of the past.

FUNGUS

Fungi may grow in any garden at any time with no signs or anything in advance. It can cause your plants serious problems, such as disease growth, rust and mildew. The symptoms of the fungus are clearly apparent as whitish patches, the leaves tend to have white powder on them, and the leaves turn brown and soft. To deal with this, there are essentially two different types of actions: non-organic fungicides described as "safe" but containing dangerous chemicals for humans and animals. And herbal home remedies' are available. Seek to get rid of fungi using any of those basic methods. Neem air is a great fungal deterrent. What you have to do is keep reapplying every other day until the problem is gone. Another brilliant idea is a combination of a coffee solution: Do 1 part of all black coffee to 10 parts of tea. Mix all of this in a spray bottle. Another solution which is known to have results is one tablespoon of baking soda per gallon of water, one dab of dish detergent and just one drop of vegetable oil. Thoroughly spray this solution over your plants.

WATER REVISITED
ADVANCED NUTRIENT MANAGEMENT FOR HYDROPONIC GROWERS CHECK YOUR NUTRIENT IQ

Nutrient management provides an opportunity to improve plant development for the professional hydroponic growers. This poses a difficulty for the beginner to deal with. The difference is in awareness, understanding and equipment. Consider the following questions for checking your nutrient IQ: What is your nutrient solution temperature, what is the range during a day and during a season?

What is the quality of the water you use to combine your nutrient with "dissolved solids" and does that quality differ significantly from season to season? Does your water supplier send you good water at one time of the year from one reservoir and poor water from another reservoir at another?

Does your water contain any components that could influence the availability of nutrients to your crop?

What is your nutrient's "EC," or strength? Can you combine specific nutrient blends for various plant types and for the life-cycle of each stage of the crop?

Will its nutrient stay within an acceptable range?

Will your nutrient contain any pathogens from a polluted water source or from sick plants that can spread disease to the rest of your crop?

Should you change your nutrient often enough to prevent salt accumulation excesses or nutrient deficiency deficiencies?

Do you know that eliminating the wastes your plants discard into the nutrient is an important reason for adjusting your nutrient solution?

Are you aware that as plants transpire, the levels of moisture and nutrients drop in your reservoir, and the EC or nutrient intensity will increase to dangerous levels?

Here are only a few simple questions that will help you to better understand what you already know, and what you may need to learn every time to produce outstanding crops.

This topic is particularly for the advanced grower, who wants to achieve the highest yields and is seriously interested in being at the forefront of plant growing technology. Hobby farmers usually don't have to think about any of these things, but just don't stop reading.

When there are issues, and a crop is not growing as well as it should. The nutrient management issue can also be traced to. If you know what can go wrong, when it happens, it's easier to identify a question. What distinguishes hydroponics from soil agriculture is root climate. Plants expect rainfall or precipitation in soil, and their roots are searching for important nutrients. It thrives with healthy, fertile soil and abundant water plants.

Through hydroponics, we continuously supplied the plant roots with water, oxygen and nutrients— no looking for the nutrients available or waiting for the next storm. The challenge for the grower is to keep up with the needs of the plants and avoid destroying plants with mineral excesses or shortages, pH and temperature levels, or a lack of oxygen. A few basic tools and strategies can make the difference between failure and performance.

• WHAT *IN THY* WATER?

The first matter to remember is the consistency of the water. It's easy to succeed with the fine, soft water. Just add the proper nutrient combinations to the water, and you're off and rising. You will have to filter your water using "reverse osmosis" if you have really hard water, or water polluted with salt, sulfide, or any number of heavy metals. So, what's in your water anyway? The most full solution is to have your water tested by a laboratory. Call the water department if you are on a municipal water system and request a copy of their new report.

Another solution-highly recommended-is to periodically test the water with a dissolved solid meter, also known as an electrical conductivity (EQ) meter or parts per million meter (PPM). These instruments are one of the most critical resources a grower requires and uses on a regular basis.

Both of these instruments work in the same way, basically. We gage the water's electrical conductivity.

In most water, it's the dissolved salts that allow it to conduct electricity. Clear water is a weak conductor since none of the conductive salts contained in impure water is present. When measuring with a meter of dissolved solids, filtered water should show no, or very low, salt content (conductivity).

Finding high salt levels in well water or municipal water sources isn't unusual. Calcium carbonates and magnesium are among the most popular ingredients in tap water and well water. In fact, water "hardness" is characterized as a measure of the calcium and magnesium carbonate content of the water, or sulfates.

As calcium and magnesium are essential nutrients for plants; for hydroponic cultivation, water with appropriate levels of these elements can just be perfect. Nonetheless, if the rates are too high even a positive thing may become a problem.

In most hydroponic applications, a calcium level of more than 200 PPM, or 75 PPM in magnesium, is usually at the verge of excessive. An excess will cause the "lockout" and become unavailable of other essential elements in the nutrient solution. For instance. Excess calcium can bind to phosphorous to produce calcium phosphate which is not very soluble and therefore not available to the crop. The aim is to start with decent water and to add the proper nutrient combination.

• TOO HOT, TOO COLD

Temperature of the water is another important factor. When your solution is too cold, seeds will not germinate, cuttings will not root, and plants will grow slowly-or cease to grow and die. If it is too dry, the same seeds will not germinate, cuttings will not grow, and plants will die as a result of oxygen deficiency or simply as a consequence of temperature stress. Many plants prefer a temperature range of root zone between 65 degrees (18 C) and 80 degrees (27 C), cooler for winter crops, warmer for tropical cultivation. It is a good idea to allow it to come to the same temperature as the water in the reservoir when adding water to your reservoir.

Notice, the plant roots have formed in a soil environment where increases in temperature occur gradually, tempered by the earth's thermal mass.

Plants do not like speedy changes in temperature, particularly in the root zone!

WATER PH

A topic frequently discussed but not understood by many growers is pH nutrients. Generally, we are concerned about it and its impact on the availability of nutrients. For example, iron may become inaccessible if it is too high. Even if your nutrient solution may have sappropriate iron content, it may not be absorbed by your plants, resulting in an iron deficiency: the leaves of the plant turn yellow and weaken.

At the other hand, advanced plant-based hydroponic foods contain unique "chelates" designed to ensure iron availability at its higher levels. It results in your crop growing fairly well. But at higher pH. Nevertheless, high pH in many ways can kill plants, the cause of a high pH solution can be very complex. Most sources of city water contain calcium carbonate to elevate it of the water and avoid corroding of pipes. As a result, you start with water that has an abnormal pH, usually 8.0 for urban water.

The begging way to deal with this is to combine fresh nutrient with your water, let it stand for a while to settle, then check and change it. City water sources also allow you to add a bit of it down (usually phosphoric acid) to lower its level for most plants, between 5.8 and 6.2. As the plants grow, checking the pH periodically is a good idea, and changing it if necessary. Without modification, you can safely let it drift between 5.5 and 7.0. In addition, continuous dumping of chemicals into your system to maintain a perfect pH of 5.8 to 6.0 will cause significant harm. Drifting up for a while, then down, then up again is normal for it. This shift is a sign your plants are adequately absorbing nutrients. Just change it if it goes too far.

A pH of less than 5.5 or greater than 7.0 will mean trouble. But don't overreact. The effect of a malfunctioning pH meter may be a seemingly unexpected and drastic change in it. If in doubt, double test before changing your solution with a reagent(colour match) pH pack.

Note also that all methods of measuring it are dependent on temperature.

CHAPTER NINE

GROWING VEGETABLES AND HERBS HYDROPONICALLY

- Constructing a drip feed system
- Maintaining a drip feed system
- Growing herbs the hydroponic way.
- Seed germination
- Plant combination
- Vegetables to grow
- Seed selection

Now that we know the fundamentals of Hydroponics, it's time to create our own hydroponic system and learn in your system to grow tomatoes! However though there are many hydroponic systems, using a' Drip System,' which is generally used to grow larger plants, we must learn to grow tomatoes.

CONSTRUCTING A DRIP-FEED SYSTEM

A drip feed system has a dripper which provides nutrients for each plant. Since the nutrient solution is fed to the roots directly, they need not expand long in search of nutrients. Sections of a drip feed system are:
1. Drip feed tank
2. Drain System
3. Pump
4. Drippers
5. Growth tray
6. Pipe

Step 1
Drill a hole at the bottom of the growth tray which holds plants and growth material. Attach a pipe to this hole which runs into the tank of nutrients. The edges of the hole are coated using silicon sealer. This pipe helps pump nutrient waste solution back to the tank. Drippers Main line Air Pump roots nutrient solution returning nutrient tube solution into reservoir.

Stage 2
Connect the pump and the main nutrient solution carrying loop.

Step 3
Now connect the dripper tubes to provide individual plant nutrient solution.

Step 4
Fill the growth tray with the material for growth (your choice) and position the plants. Enable ample space between plants.

Step 5

Turn on the pump and ensure that the correct quantity of nutrient solution is dripped from the drippers.

MAINTAINING YOUR DRIP FEED SYSTEM

- Start your drip feed system always with a full nutrient solution bottle.
- Add a gallon of pH-adjusted water before toping up your container; let your machine run for 5 minutes.
- Fully and carefully drain the system
- Fill your jar with a fresh batch of nutrient solution tailored to pH.
- Clean the drippers and the pump every 2 to 3 months.
- Clean your grow tray and air-dry after harvest.
-

GROWING HERBS THE HYDROPONIC WAY

Almost every one of us would love the thought of using freshly plucked herbs in cooking. While you can still grow these herbs in soil, if you grow Hydroponically, you can reap a healthier plant. You will find out more about growing some herbs Hydroponically on the following pages.

BASIL

The best growth propagation method is NFT-moving seedlings to NFT when seedlings are 1 to 2 inches tall. Seed germination takes 7 days; give between plants a spacing of 9 to 12 inches. Vermiculite, soilless mixtures, Rockwool and coco-peat can be used. Basil thrives in adequate sunlight and continues to expand, and requires 11 hours of daylight. Should not grow basil for short days as it is vulnerable to illness. Aphids and Pythium make up the most common causes of disease. USDA ruggedness does not refer to Basil.

CHAMOMILE

Chamomile is an annual herb with small white flowers; these flowers are added to the tea to give a distinct flavour and induce medicinal properties. Chamomile seeds need to germinate in light and typically take 7 to 14 days to germinate. This herbs grow to 20 to 30 inches long, and do well with a 6 inches spacing between plants. Aphids and Mealybugs are possible pests.

CHERVIL

This herb belongs to the family Parsley and is used mainly as a culinary herb. This herb thrives at a cool temperature (700 to 750 F); prevent it from being exposed to direct sunlight. Germination time for Chervil is around a week; when the first true leaves appear, transplant the seedlings into NFT; Spacing between plants must be kept at 1 inch, and you can start harvesting in one month. Especially during warm days, Chervil is vulnerable to Aphids.

CILANTRO

Cilantro or Coriander is a member of the Parsley family; cilantro tastes more like peregrine but with a twist of citrus. These plants produce seeds used in different mixtures of spices, liquors, and confectioneries. One may use the leaves as a culinary herb. Cilantro grows well in sunlight; Fluorescent lights or HID lights may be used. Germination of seed takes about 10 days, and the plant can grow to 2 feet tall. Aphids, whitefly, mites and thrips are the plagues that can attack cilantro. Annual USDA Hardiness.

DILL

Dill is also a member of the Parsley family, and is an annual/biennial herb. Dill seeds are used in pickles, and the herb itself is used as an enhancer of culinary tastes. Such plants typically reach 24 to 36 inches high and have to be spaced 12 to 15 inches apart. The seeds require about 10 days to germinate, and the plant grows well in full sun. Indoor use of HID lights will do the trick. Care must be taken not to overwater Dill plants; USDA hardness is not unique to Dill, and powdery mildew and aphids can damage it.

LAVENDER

This flowering plant is part of the mint family. Lavender flowers offer a good fragrance when dried; these flowers are commonly used in oils and perfumes used for aromatherapy. Lavender grows to 18 inches tall; the distance between the plants must be 20 to 24 inches. It can take 10 to 28 days for the seed to germinate. Lavender requires some sunshine; replicate the light using HID lamps. The white-fly, spider mites, Mealy bugs, and scales will invade lavender. USDA Hardiness is 5a to 9b

LEMON BALM

This is part of the mint family and is known for its soothing characteristics. It is known as a calming herb. Lemon Balm grows well under partial sunlight and propagates through the foliage and seeds. Crop germination takes 5 days, and root development takes 3 to 4 weeks. The seedlings transplant when they are around 2 inches tall. Lemon Balm is susceptible to attack by White-fly, Spider Mite, and Thrip. USDA Hardiness Range is between 4a and 9b.

MARJORAM

These are perennial herbs grown as annuals; wild and sweet, there are 2 varieties. Sweet Marjoram is used as a vegetable herb. It grows to a height of up to 36 inches, and there must be 15 to 18 inches of space between plants. Germination time is 10 to 15 days; under sunlight, those herbs grow well. Lack of light or improper lighting arrangements can cause fungal infections to perish from the plant. Marjoram is susceptible to a number of fungal infections and attacks by white-fly. USDA Reliability is 6b to 11.

OREGANO

A perennial herb and' plant marjoram' is also named. Oregano, a Mediterranean herb grows to a height of 18 inches; spacing between plants must be between 12 and 15 inches. The time of seed germination is 8 to 14 days, and it takes about 4 weeks to begin harvesting. This herb likes to grow in full sunlight; do not overload the leaves or it will cause decline and infections. When daylight is less than 11 hours, you need to have extra lighting. Oregano is susceptible to white-fly, mite and leaf-miner fungal infections and attacks. USDA Rigidity is 5a–9b.

PARSLEY

A biennial herb, grown annually. Parsley has various variations—curly leaves, fern leaves and roots. Parsley can grow in maximum sunlight, but it can also grow in partial shade. It is enough that fluorescent lamps are used for indoor cultivation. Such plants have a maximum height of 18 inches; the plants are spaced 12 inches apart. In Parsley, seed germination is sluggish-typically; it takes 3 weeks for the seedlings to grow. Parsley isn't associated with many diseases; however, Aphids, spider mites, and white-fly may attack them. USDA Rigidity is 5a–9b.

ROSEMARY

Rosemary has many uses; it's used in medicine and as a culinary herb. It herb thrives in well-drained soil, and sometimes needs watering. Rosemary has a slow rate of growth but can last with sufficient winter security for years. They can grow to 6 feet tall; germination of seed takes around 3 to 4 weeks. The rate of growth slows when the light of day falls below 11 hours. Rosemary is susceptible to fungal infections, including Powdery Mildew and a Mites attack. USDA Durability is 7a to 10b.

SAGE

Sage is a perennial shrub used for both culinary and medicinal applications. Sage requires ample sunlight, with at least 10.5 hours of daylight. Shorter days are affecting the shrub's growth. Germination of seed takes about a week, and grows up to 36 inches long. The spacing between the plants will be 24 inches apart. You can use 14 hours/day of additional fluorescent lighting during winter to ensure good growth. Sage is vulnerable to infections but is usually able to withstand them; it is resistant to attacks from Mites and White-fly. USDA Durability is 4a–11.

A NOTE ON CULTIVATING HERBS FOR COMMERCIAL USE

Herbs are often in demand–for food, medicine, aromatherapy, etc. When you start growing herbs, it is almost an addiction! The taste and fragrance of your garden's freshly plucked herbs are both satisfying and enriching. As you gain traction with herb gardening, you'll soon know the well growing herbs and those that don't. You may even get your favourites. When your garden is already full of herbs, and your neighbours are beginning to borrow fresh herbs from you, it's time to take your hobby to the next level. Some of you may be frightened by the thought of growing herbs as a company while others may be delighted. Whatever the way you feel about it, market research still has to continue.

Market research gives you useful insight into what sells and doesn't. Market research is easy to carry out. Take a notebook and a pad to your nearest grocery store or supermarket. The first step is to realize what your area's in demand for herbs. Basil is typically always common because it is heavily used in Italian cooking-remember sauce from Pesto! You can be surprised to find, when you start your study, that most fresh herbs lack freshness.

When you've done the supermarket research, visit a nearby grocery store and do another study. Here you can speak to the manager or owner directly, and speak to them about supplying their stores with fresh herbs. Of course, anyone you approach will ask for samples. While holding your items, ensure they are perfectly bundled in a ziplock bag or a clear polythene cover. You will go a long way with the way you approach things. Please put the extra effort into making your presentation appealing. Punch any holes in the plastic bag so that the herbs can breathe. Carry your samples in icebox. This keeps the herbs fresh—especially if you're travelling on a hot day with herbs sitting in the back of your car.

When you put in extra effort to sign your bags with a brand name, you can can score extra points with your prospective buyers. I say —design your herbs brand name, print labels and place it on your sample bags. This way, you can create your own brand name. It reveals you as an entrepreneur too. You can also print business cards-there are a number of inexpensive ways to do them. Check the Internet, and you can find several websites that offer free business card printing.

If you go out to meet potential customers, always take fresh samples with you. Explain the difference between herbs grown in the soil and herbs grown hydroponically to them. Tell them that the herbs that you cultivate are free of pesticides, herbicides, and toxic substances. Let your sample talk.

Once you have obtained the order to supply, it is necessary to meet the demands. Be realistic; if you do not have the facilities and resources to satisfy their demands, do not approach supermarkets. Starting with your nearest grocery store or farmers ' market is always a good idea. Second, you should sell samples that are safe for people to enjoy. As already stated, the key is good quality. When people know that the quality of your herbs is good, they would not mind charging you a slightly higher price too. Starting small is always better, and rising tall will be attained. In this way, to step up to the next level, you reduce the risk, gain experience and trust.

SEED GERMINATION

Seeds are the main vegetable reproductive source. You can also transplant soil-grown plants into your hydroponic system, but it's the easiest way to start from the seed—you don't have to worry about transplant shock and pests in your plants. To start and end the germination process successfully, there are a few criteria to be met:

• Choose a substrate—Perlite and Rockwool are the most common ones

• Moisten the soil with a nutrient solution (dilute the solution to achieve half strength)

• Maintain pH of a nutrient solution at 6.0

• Sprinkle the seeds on moistened Perlite; while using Perlite content; cover it with a moistened Perlite;

- Cut Rockwool into cubes, make a hole in the center of each cube and soak in a nutrient solution if you prefer. Now lower one seed into each cube's cavity.

- It is vitally important to maintain optimum humidity, air temperature and root temperature.

- Moisture must range from 70 to 90 percent

- Air temperature must be about 70 ° F to 78 ° F

- Root temperature must be between 72 ° F and 80 ° F

- 20W / sq. Footlight till most seeds starts sprouting.

- Increase lighting once sprouts appear

- Wait for the 2nd set of leaves to appear then move to the growing field.

- Remember to over-seed 25 to 50%-not all seeds can become strong plants; some may not germinate at all and others will die soon after germination. Overseeding lets you select the best plants to move to your area of production.

- It is critical for proper seedlings to maintain the correct temperature –this is the phase before sprouts appear. Incubators, propagation table, or seedling heat mats may be used to maintain the appropriate temperature.

PLANT COMBINATIONS

There are different combinations that you would want to consider if you intend to develop more than one crop at a time. Different plants have common needs- nutrient requirements, time for germination, growth stages, etc. Cultivating these crops together will help you achieve better quality and quantity of crops. Here's a list of crops with similar needs.

TOMATOES

Tomatoes are the most common Hydroponically grown vegetables. Some of the commercially grown tomatoes are indeterminate varieties. Select seeds accordingly if you wish to grow smaller tomatoes to fit your space requirements.

Tomatoes can be developed in any Hydroponic system, but the best is a Drip system. Crop germination time is 3 to 6 days, and it will take 100 days to see your tomato plants bear fruit; they normally keep growing fruit for one year.

LETTUCE

Peppers and Cucumbers: An all-time favourite of Hydroponic growers. Leaf lettuce is a superior alternative to head lettuce.

Nutrient Film: Technique is Lettuce's best growing system; you can grow it in an Ebb and Flow or drip feed system, too.

Lettuce seeds germinate in 4 to 8 days and can be harvested in 35 to 45 days; seed them every few days to keep a continuous supply of lettuce.

Related crops: spinach, basil and leaf crops.

CARROTS

Carrot is a well-growing root crop in perlite, just like any other root crop like radish and beets.

Roots crops require a large bed growth to grow and fully develop.

It takes the seeds 6 to 10 days to germinate, and 2.5 to 3 months to harvest.

Related crops: Beets, Leeks, Radish

CUCUMBER

Cucumbers are long-term crops and even produce, for up to 6 months, fruit. European seedless varieties are Hydroponically easier to produce. The cucumbers need large space to develop and help.

Drip system with Rockwool or Perlite as your best option for the growth material. Germination takes 3 to 5 days, and harvesting, for 6 weeks; then you will start harvesting for six months. To promote the growth of cucumbers, harvest them regularly.

BASIL

Basil peppers and tomatoes are a herb which grows between 12 and 18 inches tall. Daily pruning of buds and flowers is needed to promote continuous growth.

The Basil seeds require 6 to 10 days to germinate and will grow fresh leaves for 3 to 4 months. The old basil plant must be replaced with a new after 3 to 4 months.

Related crops: Spinach and Lettuce Beans High yields can be obtained from bean plants. With Perlite or extended clay pellets, an Ebb and Flow method works well.

Crop germination takes 3 to 8 days, and harvesting will begin in 6 to 8 weeks. The harvesting will continue for 3 to 4 months.

Related crops: Radish peas is also a close root crop to carrots. Radish grows well in a deep-seated bed system. Common growth systems are an Ebb and Flow system or a perlite or LECA drip system as growth materials.

The germination of seeds is fast-it takes 2 to 5 days. Your first harvest will start within 30 to 40 days.

BROCCOLI

Broccoli belongs to the cabbage family of Leeks, Beets and Carrots and likes to grow at cooler temperatures. So, if this vegetable is to be grown, make sure your temperature settings are cooler.

Seed germination takes 5 to 6 days at a temperature of 770 to 950 F, and it takes from 10 and 20 days at a much lower temperature of 500 and 590 F. Within around 4 months, you can start harvesting Broccoli.

Related crops: Cauliflower, Cabbage

PEPPERS

Peppers are grown in various colours and flavours. You can grow hot peppers as well as sweet one. To grow these vegetables, use a drip system, or Ebb and Flow system.

Pepper plants grow tall–it is crucial to use a growing material that provides the required support; you can use Rockwool for germinating the seeds, and also use Perlite mixed with Vermiculite and fine gravel.

Seed germination takes 10 to 14 days, and you will harvest in the 4th month.

SPINACH

Spinach tomatoes and cucumbers grow well within an NFT or Ebb and Flow system. You may use Rockwool to germinate seeds; care must be taken to provide ample growing space for these plants otherwise you risk choking them out. The average space between two plants is 20 sq centimeters.

Seed germination takes 6 to 12 days, and the first harvest only occurs at the end of the second month (approx. 50 to 60 days).

Related crops: Basil, Lettuce.

VEGETABLES TO GROW

Vegetables provide the human body with vital nutrients, and form part of a balanced diet and lifestyle. Unique vegetables can be eaten raw —as salads, while other vegetables need cooking. Whatever way we eat vegetables, they're delicious and offer tremendous health benefits.

Growing your vegetables and using them in your everyday cooking is much more fulfilling! Not only do homemade vegetables taste good, but they are much safer than commercially grown crops. Nothing works better if you are health conscious than a bowl of freshly made salad from home-grown herbs and plants.

Plants can be periodic, perennial, or biennial. Perennial plants have been working there for over two years. Annual plants germinate, bloom, and die within or within one year of the season. It takes two years for the Biennial plants to complete their life cycle.

LIST OF VEGETABLES THAT CAN BE GROWN INDOORS SUCCESSFULLY SALAD GREENS

- Spinach–germination time is 6 to 12 days; is an annual winter crop
- Watercress–seed germination occurs in 7 to 14 days; is a perennial crop
- Arugula–germinates in one week; is both annual and perennial
- Mustard–germinates in 5 to 10 days; different seed varieties sprout in one day!
- Lettuce–it takes between 4 and 8 days to germinate; is an annual or biennial temperate

HERBS

- Dill–germinates in a week period; is an annual herb
- Coriander–germination period is typically 10 days; is an annual herb
- Basil–germinates in 6 to 10 days; is an annual herb
- Chamomile–seed germination time is 7 to 14 days; is perennial but grows as an annual

- Lavender —takes 14 days or more for the appearance of the first sprouts; is perennial
- mint—germination time 7 to 14 days; is perennial mint

Choosing the best seeds for your Hydroponic Garden is critical. Here are a few points that will help you pick good quality seeds:

• Think about the time of blooming before buying seeds; some bloom early-in less than a month and others take more than a month.

• Buy your seeds from reputable shops; the seed packets typically contain all the details required to grow the plant.

• Consider environmental factors-plan to cultivate your plants outdoors with enough sunshine/shade or artificial light indoors. Find out how well the selected seeds are adapting to climate change conditions.

• Do not take the short path-the Internet road! These aren't all real.

• Seeds must not be more than three years old; they should have been stored in a dry cool position but never in a freezer

• The easiest way to purchase good quality seeds is to contact the breeder directly; this may not always be feasible, and you must finish buying seed from the reseller. Request 100 percent guarantee from the reseller—most genuine resellers offer a guarantee.

HYDROPONICS

CONCLUSION

You have made a first step in your career as a gardener or you have improved your knowledge in cultivation. It is now very evident that you can grow your crop all year round, have a mobile and a moveable fam with an equally small budget. Hydroponics is an innovation that seeks to provide food, for of livelihood and security to everyone. This book has tactically explained the business of hydroponics and has narrated the steps with which you can grow your fresh crops all year round, highlighting the efficiency of the system as well as the types. This book will serve as a beginner's guide that helps you to take control of your hyrdroponics garden, control pests and even know the kind of crops to grow. You will understand the science of temperature, pH and the importance of water.

Now it's time for the sorprise I have prepared for you.

The surprise is called... *Reality Bubble.*

Reality bubble is basically the social environment where you live, therefore this cognitive dynamics is composed of your family, your friends and also the contents you consume on internet. It's obvious that your social environment influence you, what would you think if you could choose your social environment to achieve your goals? I think it could be amazing and I can say that is absolutely possible.

The first thing to do could be watching videos on hydroponics on YouTube, reading books like this about the topic.

Furthermore, you could start meeting experts in hydroponics and new people interested in this kind of culture technique like you; with this type of people you could organize some *masterminds* groups where participants can share issues and ideas with the aim of supporting each other.

The most important thing to know about the reality bubble is that if you are in an efficient social environment, even if you have any problems, the reality bubble could help you. Always.

Glad that you have read this book. I hope you have been enjoying the guide. Do not hesitate to rate the knowledge gained from this book on Amazon.

Thank you and best wishes!

Matthew